Nanostructured Materials

This publication is a paperback edition of Advances in Chemical Engineering, V. 27.

Nanostructured Materials

Edited by
JACKIE Y. YING
Department of Chemical Engineering
Massachusetts Institute of Technology
Cambridge, Massachusetts

ACADEMIC PRESS
A Harcourt Science and Technology Company
San Diego San Francisco New York Boston London Sydney Tokyo

Cover image (paperback only): Illustration courtesy of Phoebe H. Smith and John S. Lettow.

This book is printed on acid-free paper. 

Academic Press
A Harcourt Science and Technology Company
525 B Street, Suite 1900, San Diego, California 92101-4495, USA
http://www.academicpress.com

Academic Press
Harcourt Place, 32 Jamestown Road, London NW1 7BY, UK
http://www.academicpress.com

International Standard Book Number: 0-12-008527-5 (case)
International Standard Book Number: 0-12-744451-3 (pb)

PRINTED IN THE UNITED STATES OF AMERICA
01 02 03 04 05 06 QW 9 8 7 6 5 4 3 2 1

CONTENTS

Contributors ix
Preface xi

Engineered Synthesis of Nanostructured Materials and Catalysts

William R. Moser, Josef Find, Sean C. Emerson, and Ivo M. Krausz

I. Introduction 2
II. Properties and Reactivities of Nanostructured Materials 3
A. Structure and Electronic Properties of Nanostructured Materials 4
B. Catalytic Properties of Nanostructured Materials 6
III. Progress in Synthesis Processes of Nanostructured Materials 8
A. Sol–Gel and Precipitation Technologies 9
B. Combustion Flame–Chemical Vapor Condensation Process 10
C. Gas Phase Condensation Synthesis 11
D. Reverse Micelle Synthesis 12
E. Polymer-Mediated Synthesis 14
F. Protein Microtube–Mediated Synthesis 15
G. Sonochemical Synthesis 16
IV. Engineered Synthesis of Nanostructured Catalysts 18
A. Hydrodynamic Cavitation 20
B. Experimental 23
C. Characterization of Reynolds and Throat Cavitation Numbers 25
D. Synthesis of Metal Oxide Catalysts and Supported Metals by Hydrodynamic Cavitation 27
E. Estimation of the *in Situ* Calcination Temperature in MoO_3 Synthesis . . 28
F. Hydrodynamic Cavitation Synthesis of Nanostructured Catalysts in High-Phase Purities and Varying Grain Sizes 32
G. The Introduction of Crystallographic Strain in Catalysts by Hydrodynamic Cavitation 34
H. Synthesis under Variable Fluid-Flow Conditions 39
V. Conclusions 42
References 42

Supported Nanostructured Catalysts: Metal Complexes and Metal Clusters

B. C. Gates

I. Introduction: Supported Nanostructures as Catalysts 50
II. Supported Metal Complexes—Molecular Analogues Bonded to Surfaces . . . 51
A. Preparation 52
B. Determination of Composition 53
C. Determination of Metal Oxidation State 53
D. Spectroscopic and Theoretical Characterization of Structure 54
E. Examples 54
F. Generalizations about Structure and Bonding 62
G. Generalizations about Reactivity and Catalysis 62
III. Metal Pair Sites and Triplet Sites on Supports 63
IV. Supported Metal Nanoclusters 64
A. Preparation 65
B. Structural Characterization 67
C. Examples 68
D. Catalytic Properties 70
E. Generalizations about Structure, Bonding, Reactivity, and Catalysis . . . 73
V. Supported Metal Nanoparticles 73
References 74

Nanostructured Adsorbents

Ralph T. Yang

I. Introduction 80
II. Fundamental Factors for Designing Adsorbents 81
A. Potential Energies for Adsorption 81
B. Heat of Adsorption 83
C. Effects of Adsorbate Properties on Adsorption: Polarizability (α), Dipole Moment (μ), and Quadrupole Moment (Q) 84
D. Basic Considerations for Sorbent Design 85
III. Activated Carbon, Activated Alumina, and Silica Gel 88
A. Recent Developments on Activated Carbon 91
B. Activated Alumina and Silica Gel 93
IV. MCM-41 94
V. Zeolites 96
A. Structures and Cation Sites 98
B. Unique Adsorption Properties: Anionic Oxygens and Isolated Cations . . 99
C. Interactions with Cations: Effects of Site, Charge, and Ionic Radius . . . 100
VI. π-Complexation Sorbents 108
A. π-Complexation Sorbents for Olefin–Paraffin Separations 109
B. Effects of Cation, Anion, and Substrate 112
C. Nature of the π-Complexation Bond 114

D. Olefin–Diene Separation and Purification, Aromatic and Aliphatics Separation, and Acetylene Separation . . . 117
VII. Other Sorbents and Their Unique Adsorption Properties: Carbon Nanotubes, Heteropoly Compounds, and Pillared Clays . . . 118
A. Carbon Nanotubes . . . 118
B. Heteropoly Compounds . . . 119
C. Pillared Clays . . . 120
References . . . 121

Nanophase Ceramics: The Future Orthopedic and Dental Implant Material

THOMAS J. WEBSTER

I. Introduction . . . 126
II. Mechanical Properties of Bone . . . 128
III. Bone Physiology . . . 128
A. Microarchitecture . . . 128
B. Structural Organization of the Bone Microarchitecture . . . 131
C. Chemical Composition of the Bone Matrix . . . 131
D. Cells of the Bone Tissue . . . 136
E. Bone Remodeling . . . 139
IV. The Tissue–Implant Interface . . . 140
A. Wound-Healing Response of Bone . . . 141
B. Protein Interactions with Biomaterial Surfaces . . . 141
C. Protein-Mediated Cell Adhesion on Biomaterial Surfaces . . . 143
V. Materials Currently Used as Orthopedic and Dental Implants . . . 145
A. Novel Surface Modifications of Conventional Orthopedic and Dental Implants . . . 147
VI. Next Generation of Orthopedic and Dental Implants: Nanophase Ceramics . . 148
A. Surface Properties of Nanophase Ceramics for Enhanced Orthopedic and Dental Implant Efficacy . . . 149
B. Mechanical Properties of Nanophase Ceramics for Enhanced Orthopedic/Dental Implant Efficacy . . . 156
VII. Conclusions . . . 159
References . . . 160

Fabrication, Structure, and Transport Properties of Nanowires

YU-MING LIN, MILDRED S. DRESSELHAUS, AND JACKIE Y. YING

I. Introduction . . . 168
II. Fabrication and Structural Characteristics of Nanowires . . . 168
A. Template-Assisted Synthesis . . . 169

B. Laser-Assisted Synthesis . . . 181
C. Other Synthesis Methods . . . 184
III. Theoretical Modeling of Nanowire Band Structures . . . 185
A. Band Structures of One-Dimensional Systems . . . 185
B. The Semimetal–Semiconductor Transition in Semimetallic Nanowires . . 188
IV. Transport Properties . . . 191
A. Semiclassical Model . . . 192
B. Temperature-Dependent Resistivity of Nanowires . . . 193
V. Summary . . . 198
References . . . 199

Index . . . 205
Contents of Volumes in this Serial . . . 217

CONTRIBUTORS

Numbers in parentheses indicate the pages on which the authors' contributions begin.

MILDRED S. DRESSELHAUS, *Department of Electrical Engineering and Computer Science and Department of Physics, Massachusetts Institute of Technology, Cambridge, Massachusetts 02139* (167)

SEAN C. EMERSON, *Department of Chemical Engineering, Worcester Polytechnic Institute, Worcester, Massachusetts 01609* (1)

JOSEF FIND, *Department of Chemical Engineering, Worcester Polytechnic Institute, Worcester, Massachusetts 01609* (1)

B. C. GATES, *Department of Chemical Engineering and Materials Science, University of California, Davis, California 95616* (49)

IVO M. KRAUSZ, *Department of Chemical Engineering, Worcester Polytechnic Institute, Worcester, Massachusetts 01609* (1)

YU-MING LIN, *Department of Electrical Engineering and Computer Science, Massachusetts Institute of Technology, Cambridge, Massachusetts 02139* (167)

WILLIAM R. MOSER, *Department of Chemical Engineering, Worcester Polytechnic Institute, Worcester, Massachusetts 01609* (1)

THOMAS J. WEBSTER, *Department of Biomedical Engineering, Purdue University, West Lafayette, Indiana 47907* (125)

RALPH T. YANG, *Department of Chemical Engineering, University of Michigan, Ann Arbor, Michigan 48109* (79)

JACKIE Y. YING, *Department of Chemical Engineering, Massachusetts Institute of Technology, Cambridge, Massachusetts 02139* (167)

PREFACE

Our ability to engineer novel structures at the molecular and supramolecular levels has led to unprecedented opportunities in materials design. It has fueled rapid development in nanotechnology for the past decade, leading to the creation of new materials with interesting nanometer-scale features. This volume presents the latest advances in this exciting interdisciplinary field, with contributions from chemical engineers, chemists, physicists, materials scientists, and bioengineers. It describes a "bottom-up" approach to designing nanostructured systems for a variety of chemical, physical, and biological applications.

Specifically, this volume focuses on the synthesis, processing, and structural tailoring of nanocrystalline and nanoporous materials. *Nanocrystalline materials* possess unique hybrid properties characteristic of neither the molecular nor the bulk solid-state limits and may be confined in nanometer-sized domains in one, two, or three dimensions for unusual size-dependent behavior. *Nanoporous materials,* characterized by well-defined pores or cavities in the nanometer size regime and controlled pore diameter and structure, give rise to unique molecular sieving capabilities and ultrahigh internal surface areas. Nanoporous structures also act as hosts and templates for the fabrication of quantum dots and quantum wires.

The chapters in this volume present detailed insights into the synthesis–structure–properties relationships of nanostructured materials. In particular, the catalytic and photocatalytic properties of nanoclusters and nanostructured materials with ultrahigh surface-to-volume ratio are demonstrated. The gas absorption characteristics and surface reactivity of nanoporous and nanocrystalline materials are shown for various separation and reaction processes. In addition, the structural manipulation, quantum confinement effects, transport properties, and modeling of nanocrystals and nanowires are described. The biological functionality and bioactivity of nanostructured ceramic implants are also discussed.

It is our hope that this volume illustrates the potential of nanostructured materials with multifunctionalities for a wide variety of applications. Chemical engineers, with their broad training in chemistry, processing, systems engineering, and product design, are uniquely positioned to play a pivotal role in this burgeoning field of nanotechnology. Active efforts in this research direction will impact how we tailor novel materials for areas such as catalysis and separations and how we integrate miniaturized systems such as microreactors, fuel cells, sensors, and batteries. Research in this exciting frontier will also lead to new devices for optical, electronic, magnetic, thermoelectric, and biomedical applications.

This volume is dedicated to Professor James Wei on the occasion of his 70th birthday, for his leadership in *Advances in Chemical Engineering* and his vision for our profession.

JACKIE Y. YING
VOLUME EDITOR

ENGINEERED SYNTHESIS OF NANOSTRUCTURED MATERIALS AND CATALYSTS

William R. Moser, Josef Find, Sean C. Emerson, and Ivo M. Krausz

Department of Chemical Engineering, Worcester Polytechnic Institute, Worcester, Massachusetts 01609

I. Introduction 2
II. Properties and Reactivities of Nanostructured Materials 3
A. Structure and Electronic Properties of Nanostructured Materials 4
B. Catalytic Properties of Nanostructured Materials 6
III. Progress in Synthesis Processes of Nanostructured Materials 8
A. Sol–Gel and Precipitation Technologies 9
B. Combustion Flame–Chemical Vapor Condensation Process 10
C. Gas Phase Condensation Synthesis 11
D. Reverse Micelle Synthesis 12
E. Polymer-Mediated Synthesis 14
F. Protein Microtube–Mediated Synthesis 15
G. Sonochemical Synthesis 16
IV. Engineered Synthesis of Nanostructured Catalysts 18
A. Hydrodynamic Cavitation 20
B. Experimental 23
C. Characterization of Reynolds and Throat Cavitation Numbers 25
D. Synthesis of Metal Oxide Catalysts and Supported Metals by Hydrodynamic Cavitation 27
E. Estimation of the *in Situ* Calcination Temperature in MoO_3 Synthesis 28
F. Hydrodynamic Cavitation Synthesis of Nanostructured Catalysts in High-Phase Purities and Varying Grain Sizes 32
G. The Introduction of Crystallographic Strain in Catalysts by Hydrodynamic Cavitation 34
H. Synthesis under Variable Fluid-Flow Conditions 39
V. Conclusions 42
References 42

0065-2377/01 $35.00

The unusual structural and electronic properties of nanostructured materials, as they relate to recent observations of enhanced catalytic reactivities, are examined. The data suggest a need for advanced synthetic nanostructured materials processes producing high-purity crystallites in variable grain sizes in the range 1–20 nm. Within the past 5 years a wide variety of new processes have been discovered, and their capabilities are compared. One novel approach has been centered on adjusting the fluid dynamics of fast-flowing, liquid-slurry streams to conduct nanostructured materials synthesis within a cavitating bubble zone. The results of the synthesis method is to produce high shear and in situ calcination by shock wave heating and high Reynolds numbers. Nanostructured materials produced by hydrodynamic cavitation resulted in exceptionally high phase purity. In most cases, finished nanostructured metal oxides flowed directly from the processor, requiring little or no postsynthesis calcination. In many cases, the grain sizes of metal oxides and supported metals could be adjusted in the range 1–20 nm, and crystallographic strain could be systematically introduced in several classes of materials such as titania, CuO in Cu Zn Al O methanol synthesis catalysts, and piezoelectric materials. The method is inherently a high-volume, continuous process.

I. Introduction

Both the discovery of new synthesis processes for nanostructured materials and the demonstration of the highly reactive properties of these materials have increased rapidly within recent years. The new synthesis processes have made available nanostructured materials in a wide variety of compositions of metal oxides and metals supported on metal oxides, which have led to recognition of their exceptional chemical, physical, and electronic properties. The objective of this review is to provide recent results on synthesis of nanostructured materials using the novel processes that were developed in these laboratories recently and to contrast them to other important, new methods. Because some of the most important applications of nanostructured materials are as catalysts for chemical processing, several key reports on enhanced catalytic reactivity of nanostructured grains will be discussed along with the pertinent theory responsible for controlling both activity and selectivity of these new catalysts.

Over the past 30 years, our laboratory has examined the synthesis of nanostructured materials for advanced catalysts. The first process for the synthesis of fine grains of cobalt molybdates for hydrodesulfurization catalysis

(Moser, 1977, 1978a) involved a supercritical fluid synthesis of a special form of alumina having all octahedral Al^{3+} on its surface. This permitted the facile growth of Co and Mo onto this surface to provide a bimetallic structure. Using similar techniques, this type of alumina was combined with a wide variety of rare earth metal ions for HDS catalysis (Moser, 1978b,c), in which the dispersion of the rare earth oxide over the nanostructured metal oxide support was high. The next catalyst synthesis process developed was the high-temperature aerosol decomposition (HTAD) process (Moser, 1991, 1993; Moser and Cnossen, 1992; Moser and Connolly, 1996; Moser *et al.*, 1993, 1994, 1996a; Moser and Lennhoff, 1984, 1989) and was based on an early patent by Ebner (1939, 1951, 1953) used for the synthesis of simple metal oxides for ceramics applications. This process had the advantage of synthesizing nanostructured grains of simple and complex metal oxides but also resulted in high-phase purities of multimetallic perovskite, spinel, and scheelite materials. Furthermore, it resulted in metastable nonthermodynamic phases due to the fine crystallite grains in higher alcohol synthesis catalysts (Moser and Connolly, 1996) and maleic anhydride, V–P–O, catalysts (Michalakos *et al.*, 1995; Moser, 1996). High-powered ultrasound was used in the synthesis (Emerson *et al.*, 1998) of a wide variety of metal oxides and supported metal catalysts. The high-shear environment and shock wave heating provided by acoustically generated cavitation resulted in catalyst grains that were in all cases smaller than classically prepared materials, and the method resulted in Au–Pt alloys supported on titania in much higher alloy phase purity (Emerson *et al.*, 1998) compared to classical synthesis. In recent years, we discovered that hydrodynamic cavitation generated by mechanical means (Moser, 1995a,b; Moser *et al.*, 1995b, 1996b; Sunstrom *et al.*, 1996) could be used to synthesize advanced catalysts, ceramics, and electronic materials in high-phase purities and as nanostructured grains. The most recent process discovered in our labs utilized a new device for the mechanical generation of cavitation (Moser *et al.*, 1999), in which the bubble dynamics, shock wave, and shear could be controlled over an exceptionally wide range using equipment manufactured by Five Star Technologies and invented by Kozyuk (1996, 1998, 1999a–d, 2000a,b). This equipment offers the opportunity to use a range of engineering techniques and fluid dynamics phenomena to control the synthesis of nanostructured materials. This is the principal emphasis of this report and is described in Section IV.A.

II. Properties and Reactivities of Nanostructured Materials

An examination of recent information on the properties, structures, and reactivities of nanostructured materials indicates the importance of discovering new synthetic procedures to fabricate such materials. Experimental data on the properties and catalytic reactivities of nanostructured materials reported within the past 5 to 10 years, coupled with theoretical computations in the older literature, demonstrate the significance of nanostructured materials having primary, crystallographic grain sizes in the range 1–20 nm. To put this discussion of novel synthetic techniques into perspective, a short review of the properties and reactivities of nanostructured materials will be given.

A. Structure and Electronic Properties of Nanostructured Materials

Although Baetzold's (1971, 1973) molecular orbital calculation performed nearly 30 years ago predicted that the ionization potential of nanostructured metals should increase as their grain sizes decrease, only recently has this been verified experimentally. Measurements on several metals having cluster sizes ranging from 2 to 140 atoms (de Heer, 1993) showed a systematic increase in the ionization potential for all metals studied as the cluster size decreased. The ionization potential of a 140-metal-atom cluster of Al increased from 4.7 to 6.5 eV for a 5-atom cluster. In our view this is an exceptionally important finding because it suggests that by systematically changing the grain size of a metal or supported metal through synthesis, one should be able to adjust the overlap integral between the donating metal electrons contained in nonbonding orbitals and the empty antibonding orbitals of a substrate so that an optimum energy relation exists for electron transfer and activation of either a catalytic or noncatalytic reaction. The basic theory for noncatalytic systems was described by Woodward and Hoffmann (1970), and the relationship of metallic band theory to reacting substrates was described by van Santen and Niemantsverdriet (1995). The data and reaction theory suggest that a key requirement of any nanostructured materials synthesis process is the capability to systematically vary the grain size of the crystallites between 1 and 20 nm without changing the chemistry by which the particles are produced. If this can be accomplished, the ionization potential of the different nano-sized grains can be adjusted for most catalytic reactions for optimum metal–substrate activation and maximized rates.

The high metal atom surface-to-volume ratio observed in nanostructured materials not only has importance to the number of active sites in a catalyst, but also can influence the oxygen and other anion-defect chemistry and the observation of metastable phases. Siegel's (1991, 1994) computations indicated that the percentage of metal atoms on the surface of a crystallite increased from a few percent in a 100-nm particle to about 90% in a

1-nm grain. The importance of this observation is that it predicts that surface free energy should dominate phase formation in nanometer grains of crystals rather than bulk thermodynamics. It predicts that metastable phases should be observed for many nanostructured materials. Our own aerosol synthesis (Moser and Connolly, 1996) of spinels resulted in the formation of nanostructured, metastable grains of a cubic spinel over the entire substitution series from $x = 0.0$ to 0.90, $Cu_xZn_{1-x}Cr_2O_4$, instead of the tetragonal phase. The synthesis of zirconia by hydrodynamic cavitation (Moser *et al.*, 1995b) resulted in fine crystallites of cubic zirconia that transformed to the monoclinic phase upon heating. Furthermore, the aerosol synthesis (Moser and Cnossen, 1992) of both MoO_3 and Bi_2O_3 resulted in the formation of metastable β-MoO_3 and β-Bi_2O_3 as nanostructured grains, which transformed to their respective α-structures upon calcination and grain growth. This aspect of nanostructured catalyst synthesis offers the possibility for the formation of high-temperature stable, metastable compositions for application in high-temperature turbine combustion. The high surface atom–to-volume ratio of nanocrystalline ceria was suggested (Tschöpe *et al.*, 1996) to be responsible for its high oxygen defect concentration and surface conductivity. This aspect is especially important to hydrocarbon partial oxidation catalysis because it affords the high oxygen mobility to balance the surface oxidation and reduction processes required for selective catalysis, but it also provides defect coordination sites for reacting molecules..

The change in grain size of a nanostructured catalyst results in the formation of cluster structures having stabilities that vary with grain size (Allpress and Sanders, 1970; Henry *et al.*, 1997; Montejano-Carrizales and Moran-Lopez, 1992). The consequence of this variation is that the fraction of (111) and (100) crystallographic planes varies according to the fraction of icosahedron or truncated octahedral phases of supported metals. Naturally, these crystallographic planes show very different catalytic reactivities, depending on the reaction under consideration. A more important consequence of the decrease in grain size is the observation by van Hardveld and Hartog (1972) that as the crystallite size of a material decreases, the fraction of atoms in coordinatively unsaturated edge sites relative to atoms in the stable basal planes increases to dominate the atoms exposed to a reacting gas. Indeed, in propylene selective oxidation it was shown that as the crystallite size of MoO_3 was systematically decreased, exposing more of the (100) crystallographic edges, both the activity and the selectivity for forming acrolein increased (Volta *et al.*, 1979; Volta and Tatibouet, 1985). However, for some catalytic processes, a small grain size can work to the detriment of selectivity. Okuhara and co-workers (1993) showed that exposing more of the edge sites of a V–P–O catalyst in the direct oxidation of butane to maleic anhydride resulted in a sharp decrease in selectivity.

B. Catalytic Properties of Nanostructured Materials

The preceding discussion points to a strong relationship between catalytic activity and selectivity that is controlled by grain-size effects. Although many studies, which are summarized later, have clearly demonstrated that turnover numbers (TON) greatly increase as the grain sizes decrease, the factors responsible for the observed rate acceleration have not been elucidated. The preceding discussion suggests that some of the controlling factors responsible for rate acceleration are (1) the increase in coordinatively unsaturated edge sites as grain sizes decrease, (2) the more favorable energy relationship between electrons on the metal or metal oxide surface and the antibonding orbitals of the substrate to be activated, (3) a higher concentration of anion vacancies and defect sites on the crystallite surface, (4) the higher electrical conductivity and ion mobility of nano-crystallite surfaces, (5) a higher degree of crystallographic strain as crystallites become smaller, and (6) alteration of the concentration of crystallographic planes of greatly differing activities as the catalyst crystallite size changes in the region 1–20 nm. In the following section, a few examples of a decrease in a catalyst's grain size leading to increased reactivity per metal atom center are given; however, there are some cases reporting that a decrease in the grain size resulted in a volcano plot, where reactivity decreased as the particle approached the 1- to 3-nm level. This phenomenon is expected if there is quantum mechanical control over the reactivity or when the particle becomes so small that it reacts with the support surface to generate a less reactive metal center.

The literature reports several cases where unusual catalytic properties resulted when nanostructured materials were compared to micrometer-size grains. When 1- to 5-nm particles of rhodium were synthesized on a polymer support (Busser *et al.,* 1996) and studied for their activity to hydrogenate cyclohexene, the turnover number smoothly increased by a factor of 10 as the particle size decreased from 5 to 1 nm. The synthesis of TiO_2 as nanometer grains by an aerosol process (Wold *et al.,* 1996) resulted in greatly enhanced activity as a photocatalyst for the decomposition of chlorocarbons. Thin films of TiO_2 were fabricated by spray pyrolysis techniques and were found to be especially active for chloro-organics photodecomposition. A sample of MoS_2 prepared with a high edge-to–basal plane ratio as 5- to 25-nm crystallites by an exfoliation technique led to a hydrodesulfurization activity that was four times greater than that of a large crystal of MoS_2 (Del Vallee *et al.,* 1994). Nanometer-size grains of supported iron and palladium catalysts (Wilcoxon *et al.,* 1993b) prepared by a micelle technique showed that pyrene was hydrogenated progressively faster up to two orders of magnitude in rate ratios when the particle sizes of palladium were systematically decreased from 14 to 3 nm. The preparation of gold particles in different grain sizes on titania and iron oxide supports led to catalysts that were substantially more active (Haruta, 1997; Haruta *et al.*, 1993; Sanchez *et al.*, 1997; Tsubota *et al.*, 1994, 1995) and more stable for CO oxidation. The gold particles were found to be in a separate phase from the support and were a few nanometers in grain size. As the grain size was reduced from 8 nm to the 3-nm range, the reactivity increased dramatically. Other Au studies (Sze *et al.*, 1993) on CO oxidation on iron oxide of different morphologies showed that both grain size and the degree of roughness of the catalyst were important in controlling the CO oxidation activity. Grain sizes of 2–6 nm exhibited nearly twice the rate of 3- to 8-nm particles, and very uniform Au particles demonstrated a very low rate. In addition, the nonuniform Au particles deactivated at a much lower rate than uniform ones. Suslick and co-workers (Grinstaff *et al.*, 1992; Suslick *et al.*, 1994) reported the synthesis of nanometer grains of iron particles prepared by acoustic sonochemical means. These catalysts were prepared by sonolysis of $Fe(CO)_5$ in high surface areas as amorphous metallic particles due to the high heat-up and cooling rates inherent in the cavitation effect. They were much more effective than classically prepared materials in syngas conversion and hydrocarbon dehydrogenation.

Studies on the formation of metal clusters between 1 and 50 atoms by ion-bombardment techniques (Kaldor and Cox, 1990) led to several interesting phenomena as the number of atoms in the clusters decreased into the range 3–15 atoms. Deuterium adsorption studies showed that the H/M ratio greatly increased as the number of metal atoms per cluster decreased into the 10-atom range and greatly accelerated between 3 and 5 atoms for Rh, Pt, and Ni clusters. Similar preparations of Pt and Pd clusters were evaluated for their reactivity with methane (Cox *et al.*, 1990). Pt clusters showed a marked increase in reactivity below 8 atoms per cluster, whereas the Pd clusters exhibited a maxima between 10 and 15 atoms. Doesburg and co-workers (1987) synthesized copper supported on alumina-stabilized zinc oxide in grain sizes ranging from 3 to 20 nm. When these catalysts were evaluated for their methanol synthesis activity, a smooth increase in their turnover numbers resulted as the grain size decreased but showed a maxima at 4.5 nm. Smaller grain sizes resulted in much lower reactivities. Such a decline in activity for very small particles was observed in butane hydrogenolysis and hydrogen adsorption by nanostructured Pt supported on fine grains of titania (Salama *et al.*, 1993). The studies showed that the platinum was better dispersed as the titania grain size decreased, and the catalytic reactivity increased at lower Pt grain sizes down to 4 nm. Electron spin resonance studies showed that the reason for the reduction in activity for Pt below 4 nm was due to an SMSI surface reaction of Pt(0) with titania to form a partially oxidized Pt atom and formation of surface Ti^{3+}. Likewise, the hydrogenation of benzene by Rh supported on alumina (Fuentes and

Figueras, 1980) in grain sizes from 11 to 1 nm showed that as the Rh grains decreased from 11 to 4 nm, the TON increased, but it then sharply decreased from 4 to 1 nm. Benzene hydrogenation on Ni supported on MgO showed no such volcano plot (Nikolajenko *et al.,* 1963). Although the Ni grain sizes were not measured, the hydrogenation rate per metal atom in the samples smoothly increased by a factor of 20 as the Ni concentration was decreased from 80 to 1%. Masson *et al.* (1986) reported that ethylene hydrogenation by Pt on alumina exhibited an increase in TON by a factor of 6 as the grain size of Pt decreased from 2.5 to 0.5 nm, which decreased by a factor of 2 at slightly lower grain sizes.

Our conclusion from these properties and reactivity considerations is that the discovery of nanostructured synthesis processes that afford systematically varying grains in the range 1–20 nm is exceptionally important to the discovery of highly active and selective catalysts. It is evident from the general catalysis literature that if one is to be able to observe systematic changes in catalytic performance as a function of the degree of ion modification, one must use synthetic techniques that ensure a high degree of phase purity. This requires a high degree of mixing of ions during the synthesis step or avoidance of phase separation during this step. Naturally, the optimum situation is to achieve homogeneous solid solutions after calcination to the temperature where the catalytic process operates. Fortunately, most nanostructured synthesis processes produce materials having high phase purities, whereas most classical methods of synthesis result in rather poor phase purities.

III. Progress in Synthesis Processes of Nanostructured Materials

Due to the recent identification of the unusual reactivities and structural properties of nanostructured materials, progress the syntheses of such materials by entirely new approaches and modifications of older methods has greatly accelerated. The older methods have recently produced nanostructured grains (smallest crystallites) and particles (agglomerates of fine crystals) in controlled morphologies due to a concentration on the chemistry required for the formation of fine grain materials. Many of the newer processes have required the development of special hardware and equipment to give particle isolation through metal vaporization or molecule vaporization, metal ion dispersion and mixing through inducing a high shear environment during synthesis, atomic scale separation of the metallic components through surfactant coordination or supercritical fluid processing, or the introduction of low concentrations of mutually insoluble solid-state phases during the synthesis step. Many of the more important methods are described in the following sections.

A. Sol–Gel and Precipitation Technologies

Continuing progress is being made in the synthesis of nanostructured materials by sol–gel techniques in the synthesis of nanolaminates (Sellinger *et al.,* 1998) and solution-precipitation techniques (Schultz and Matijevic, 1998). The advantage of the sol–gel method of synthesis is that virtually any metal oxide system can be examined, and no special apparatus or equipment is required. The advantages of sol–gel processing are the formation of ceramics of high purity and good control over microstructure and particle morphology in the synthesis, typically at room temperature. Matijevic (1992) has shown that the precipitation of colloids may be accomplished in a wide variety of morphologies and grain sizes; however, in order to obtain different morphologies and sizes, extensive chemical modifications must be made for each metal oxide system studied. The sol–gel technique works well for the synthesis of complex metal oxides with high phase purity because the polymerizing gel traps the various metal ion components spatially, permitting precipitation from solution where all the metal ions occupy near-neighbor positions in the gel matrix. Upon further processing and high-temperature calcination, the resultant amorphous mixture of metal oxides, hydroxides, and metal salts decomposes with M–O–M bond formation, while having to diffuse only a few angstroms to their lattice positions in a homogeneous solid solution. Furthermore, the gel matrix isolates the individual metal oxide particles, giving rise to nanostructured grains after the high-temperature calcination to form the finished metal oxide. Other methods use organometallic precursors having well-defined numbers of metal ions in their organometallic clusters as either mono or multimetallic complexes.

A study (Zhang *et al.,* 1998) on the synthesis of nanostructured titania reported that the introduction of low concentrations of iron during the precipitation step resulted in anatase having photocatalytic activity considerably higher than that of materials containing no Fe. Samples were prepared in grain sizes of 6, 11, and 21 nm by carrying out the hydrothermal synthesis at different temperatures and times, and the variation of Fe between 0.0 and 1.0 resulted in minor changes in grain sizes. The study further showed that despite the pure anatase crystal structure, the photonic efficiencies for the photocatalyzed decomposition of $CHCl_3$ were higher than those of standard catalysts, and there was an optimum iron concentration for maximum photocatalytic reactivity for each titania grain size. Furthermore, the photonic efficiency for pure anatase catalysts having no Fe ion doping with grain sizes between 6 and 21 nm afforded a reactivity maxima at near 11 nm.

A careful and systematic study (Wang and Ying, 1999) on titania synthesis showed that adjusting the water/alkoxide ratio in a classical precipitation of titania resulted in pure anatase materials after calcination at 723 K for 2 h, which exhibited grain sizes varying from 82 to 14 nm as the water/alkoxide ratio was systematically increased from 3.3 to 165. A modification of the classical precipitation process by precipitation followed by hydrothermal treatment at different temperatures resulted in pure anatase, which varied in grain size from 6 to 28 nm as the hydrothermal treatment temperature was increased from 353 to 513 K. Surprisingly, the 353 K hydrothermal treatment resulted in a material that remained pure anatase phase after calcination in air at 1073 K for 2 h. The sample experienced a modest grain growth from 10 to 35 nm. Performing the hydrothermal treatment in an acidic media made it possible to convert the samples to pure rutile, although the grains grew to 49 nm.

B. Combustion Flame–Chemical Vapor Condensation Process

The CF-CVC (combustion flame–chemical vapor condensation) process developed by Kear and co-workers (Skandan *et al.*, 1996; Tompa *et al.*, 1999) is a continuous process using the equipment shown in Fig. 1. The starting materials are metal complexes that can be vaporized and fed into a flat flame, which immediately converts the compounds to nanostructured metal oxides. The particle dilution is controlled to prevent agglomeration in a hot state

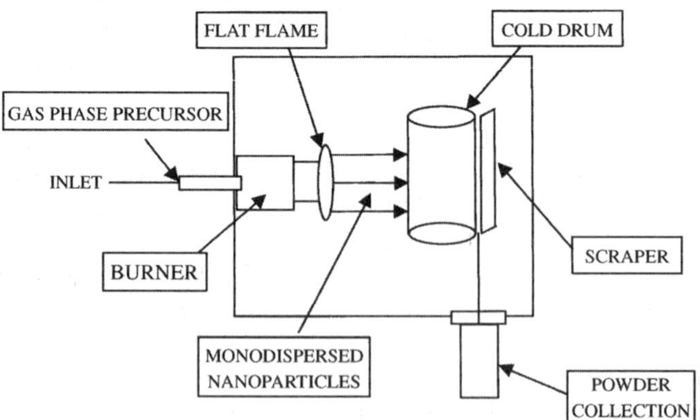

Fig. 1. The CF-CVC (combustion flame chemical vapor condensation) (Skandan *et al.*, 1996; Tompa *et al.*, 1999) for producing nanostructured materials.

in the flowing gas stream. The solid particles are then collected on a cold surface and scraped off continuously into a powder bin. A variety of simple and complex metal oxides have been produced in this way, including silica, alumina, titania, barium titanate, zinc, and copper oxides. This process is similar to the earlier-reported Cabot process (Jordan *et al.,* 1990), but it uses advanced flame design and control to afford better dispersion of the nanostructured grains. Particle sizes were generally in the 20-nm size range, although they could be controlled between 3 and 75 nm for some systems through variation of temperature, precursor flow rates, and pressure in the chamber.

The Cabot flame process (Jordan *et al.,* 1990) for the production of high-surface-area nanostructured materials is mentioned here because the technology is of great potential importance to the general synthesis of nanostructured materials. Although the technology is not current and has apparently been abandoned by the developers, we regard the approach as being exceptionally important and not generally recognized in the literature. The process atomizes a solution of a decomposable metal salt or metal salt mixture by mixing with a high-Mach-number gas at high temperature. The gas is created by high-temperature combustion of an appropriate fuel using technology developed by Cabot in other processes. The data reported for alumina synthesis indicated that high Mach numbers led to 165 m^2/g foamed alumina. The surface area of the alumina could be systematically varied between 17 to 165 m^2/g by altering the gas velocity and combustor temperature. The report also described the flame synthesis of titania and zirconia.

C. Gas Phase Condensation Synthesis

The newest development in the gas phase condensation process involved the use of dc- and rf-magnetron sputtering (Tschöpe *et al.,* 1997; Ying and Tschöpe, 1996) to introduce metals into the gas phase instead of induction heating used by Gleiter (Birringer *et al.,* 1984; Gleiter 1981, 1989) and Siegel (1991, 1994; Siegel and Hahn, 1987). The equipment used in this type of process is similar to that shown in Fig. 1 except that the starting material is usually a pure metal that is introduced into the vapor phase by various heating methods. The metal vapor is then transported by an assisted gas stream onto a cold surface where it is collected. The resultant powder may be used as is or subjected to a controlled oxidation to produce the nanostructured metal oxides. An unusually high degree of oxygen nonstoichiometry was observed in the vapor-phase synthesis (Tschöpe *et al.,* 1996, 1997; Ying and Tschöpe, 1996) of CeO_{2-x} compared to conventional precipitation synthesis. Also the grain sizes were smaller (8 nm vs 8–12 nm), and surface areas were larger than those of conventional materials. The catalytic properties of the cerium oxide nanocrystals were investigated for SO_2 reduction, where a considerable reduction in light-off temperature was observed (Tschöpe *et al.*, 1997).

Berthet *et al.* (2000) examined the deposition of palladium on a silicon carbide support (SiC) using both a high-frequency (100-MHz) plasma-sputtering technique and a metal atom beam deposition in which the Pd was evaporated onto the SiC surface. The former technique resulted in the deposition of 2.5-nm spherical crystallites on the surface, whereas the Pd atom vaporization resulted in the formation of thin stacked monolayers on the surface. In both cases, based on X-ray photoelectron spectroscopy, the low-level metal-deposited samples showed an interaction of Pd with the SiC surface. Butadiene hydrogenation using these catalysts showed that as the metal concentration on the surface increased, the turnover frequencies (TOF) increased. The synthesis studies were carried out to generate a nanostructured catalyst particle that is in close proximity to a SiC surface, an excellent heat-transporting material. The idea was to determine whether the fine-grain Pd–SiC would be beneficial in removing local heat generated in strongly exothermic hydrogenation reaction. The catalytic reactivity data showed that the thin monolayer structure (atom vapor) rather than the larger three-dimensional crystallites (plasma) resulted in far more active catalysts.

D. Reverse Micelle Synthesis

The reverse micelle approach continues to grow in importance in the synthesis of nanostructured materials. The basis for the technique is the use of a surfactant to stabilize varying aqueous droplet sizes in a hydrocarbon medium. Metal salt precursors are contained in the aqueous portion and are transformed by a reactant from the hydrocarbon phase. Both the structure of the surfactant and the steric size are able to produce metals having a wide range of grain sizes. Although notable grain-size control has been observed, the technique has a major disadvantage in that the commercial application may be limited due to the large amount of organic surfactant surrounding the crystals, and when this is removed by washing, the nanostructured metal grains normally grow.

Poly(oxyethylene)nonylphenyl ether (Igepal-20) has been used in a combined synthesis of nanometer grains of silver on fine-grain silica (Li *et al.*, 1999b). This method uses Igepal in a hexane solution with silver nitrate dissolved in the aqueous phase. Hydrazine hydrate is then added to reduce the Ag after the fine droplets of micelle have been formed, resulting in metallic silver particles contained inside the micelle. Then tetraethylorthosilicate

(TEOS) is added in cyclohexane: it is slowly hydrolyzed by the water inside the micelle, resulting in nanometer-sized grains of Ag supported on monodispersed spherical grains of silica. The advantage of the technique is that the grain size of the Ag can be systematically changed from 12 to 125 nm by altering the water/Igepal or water/TEOS ratio.

Martino and co-workers at Sandia National Lab have developed a new wrinkle in their previously reported (Martino *et al.*, 1995, 1997; Wilcoxon *et al.*, 1993a) inverse micelle synthesis of nanostructured reduced metals. Those studies showed that nanometer-size grains of supported iron and palladium catalysts could be prepared in varying grain sizes by a normal inverse micelle technique and that nanostructured palladium hydrogenated pyrene progressively faster, up to two orders of magnitude in rate ratios, when the grain sizes were systematically decreased from 14 to 3 nm (Wilcoxon *et al.*, 1993a). The new synthetic method (Martino *et al.*, 1999; Sault *et al.*, 2000) combines the inverse micelle technique with a standard sol–gel synthesis. For the synthesis of nanostructured grains of Pt/silica, an inverse micelle is formed using a surfactant in hexane–water. Then $PtCl_2$ and TEOS are mixed in, and the resultant reverse micelle is reacted with $LiBH_4$ in tetrahydrofuran (THF) to reduce the metal ion. This mixture is gelled using tetrabutylammonium hydroxide (TBAOH) and further processed by sol–gel techniques. Both Pt/silica and Pt/alumina were synthesized in grain sizes around 2 nm before calcination and 3.5–4.5 after a 723 K air calcination. Another report suggests that the technique can be used for other reduced nanostructured metals (Martino *et al.*, 1998).

A detailed study (Zarur *et al.*, 2000) of the reaction variables of a reverse microemulsion synthesis of barium hexaaluminate (BHA) resulted in nanostructured grains that could be varied over the range of 5 to 120 nm after calcination to 723 K and surface areas that ranged from 20 to 100 m^2/g after air calcination to 1573 K. The variability was obtained by altering the water concentration in a synthesis using a nonionic surfactant of polyethoxylated alcohols and barium and aluminum alkoxides as precursors. A narrower range of grain sizes of 4 to 12 nm was realized by altering the water-to-alkoxide ratio, and this variability led to surface areas after the 773 K calcination of up to 600 m^2/g. Furthermore, ceria in concentrations up to 25% w/w could be deposited on the aluminate without significant agglomeration after 1573 K calcination (Zarur and Ying, 2000). The ceria grains after calcination at 1073 and 1573 K were 6 and 20 nm, respectively. An examination of the methane total catalytic combustion activity of these catalysts showed (Zarur and Ying, 2000) that the BHA prepared by the reverse micelle technique resulted in a 110 K lower light-off temperature than a classically prepared sol–gel synthesis, and the ceria-modified reverse micelle BHA afforded a catalyst with a light-off temperature about 300 K lower.

E. Polymer-Mediated Synthesis

A potentially important, new nanostructured metal-particle synthesis technique based on mediation by polymeric materials might overcome the problems associated with the large amounts of surfactants required in micelle processes. The general synthetic technique is based on the reduction of a platinum salt (Zhao and Crooks, 1999a) or palladium salt (Zhao and Crooks, 1999b) with aqueous $NaBH_4$ in the presence of a hydroxyl-terminated dendrimer. Dendrimers are monodispersed, hyperbranched polymers having a very high concentration of surface functional groups. Twelve to 60 Pt^{2+} atoms can be loaded into a poly(amidoamine) dendrimer (PAMAM) and reduced to the metallic nanostructured grains. The interesting aspect about this method of synthesis is that the reduced metal could be extracted in to hydrocarbon solution, leaving the polymer behind. Nanostructured metals prepared by this technique have been found effective as catalysts for electrolytic O_2 reduction and allyl alcohol hydrogenation. In a review of the most recent findings Crooks and co-workers (2000) described a wide range of reduced metal-loaded particles in varying particle sizes for dendrimers containing Cu, Pt, Pd, Ru, Ni, and Ag. Depending on the degree of growth in the polymerization process, reduced metals of different cluster sizes could be formed in dendrimer formation. Furthermore, different cluster sizes could be produced by altering the amount of metal originally used in forming the dendrimer–metal complex. For example, the synthesis of Pt/dendrimers containing 40 and 60 Pt atoms within the internal structure of the dendrimer resulted in particle diameters of 1.4 and 1.6 nm, respectively. Characterization of these particles as well as Pd/dendrimers showed that the particles were monodispersed and spherical. Reduced silver particles could not be prepared directly but were obtained through *in situ* reduction of Ag^+ in solution by a reduced Cu(0)/dendrimer. Either displacement methods or direct synthesis techniques were successful in synthesizing bimetallic reduced-metal particles within dendrimers containing Au/Cu, Pd/Cu, Pt/Cu, and Pt/Pd nanoparticles. Catalytic studies for heterogeneous oxygen catalyzed reduction by 60 Pt-atom-dendrimer nanoparticles resulted in facile O_2 reduction. Due to the differing steric properties in dendrimers under different polymerization conditions, different rates of hydrogenation and selectivities between linear and branched olefins were observed. The 40-atom Pd/dendrimer catalysts resulted in turnover rates similar to those of polymer-bound Rh^+ catalysts, and when the 40-atom Pd catalyst was used after preparation from dendrimers of higher degrees of polymerization, a selectivity of the linear to branched olefins was observed. The previously mentioned catalytic studies were carried out in aqueous solutions; however, several techniques were developed to generate catalysts that were stable and catalytically active in organic media. Upon functionalization by perfluorinated polyethers, Pd nanoparticles within dendrimers could be solubilized in supercritical CO_2. Catalytic studies under supercritical conditions resulted in a much more selective catalyst for the Heck coupling of arylhalides and methacrylates, as compared to the homogeneous Pd-catalyzed reaction. One potential problem with this type of nanostructured catalyst synthesis is the fact that the polyamidoamine dendrimers have a maximum thermal stability of only 373 K, whereas the polypropyleneimine dendrimers are stable to 743 K.

Another polymer-mediated synthesis used a lyotropic liquid-crystal (LLC) template to obtain 4- to 7-nm crystallites of Pd (Ding and Gin, 2000). In this study, Pd^{2+} was ion exchanged into the channels of a cross-linked LLC and reduced with H_2. The resultant catalysts were found to be active for efficient hydrogenation of benzaldehyde.

F. Protein Microtube–Mediated Synthesis

Behrens and co-workers (1999, 2000) are developing a new process for the synthesis of nanostructured metals and alloys in varying grain sizes supported on hollow, highly oriented protein templates. The technique uses the protein α,β-tubulin after a self-assembly into 25-nm-diameter microtubes that are several micrometers long. The surface of the protein contains thiol and ammonium end groups, which are reacted in aqueous solution with Na_2PdCl_4, followed by citrate reduction. A schematic of the process is given in Fig. 2. The grain size of the palladium synthesized in this way was reported

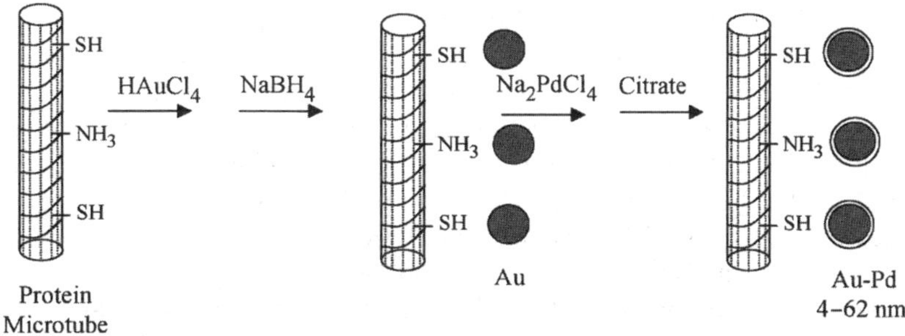

Fig. 2. Schematic of the synthesis of nanostructured metals and alloys in varying grain sizes supported on hollow, highly oriented protein templates (Behrens, Dinjus, and Unger, 1999; Behrens *et al.*, 2000).

to be 2 nm. A similar synthesis resulted in 2-nm Au particles. Upon treatment of the microtubes with $HAuCl_4$, followed by $NaBH_4$ reduction, further reaction with Na_2PdCl_4 followed by citrate reduction led to nanostructured Au/Pd alloys. By increasing the Pd/Au ratio from 2.3 to 10, the grain size of the alloy could be systematically reduced from 62 to 4 nm. The application of the alloys in crotonic acid hydrogenation in an aqueous suspension resulted in a systematic increase in the reactivity as the grain size decreased from 62 to 4 nm. The TEM data on the Au, Pd, and Au/Pd alloy particles, supported on the microtubes, appear as nanostructured grains attached to the fibrous tubes of several microns in length. Because the proteins are not stable at high temperatures, techniques were developed to support the preformed metal microtubes on alumina, which resulted in greater stability.

G. Sonochemical Synthesis

Suslick's group has made many contributions to the application of sonochemistry to the synthesis of metallic and bimetallic catalysts, which they have reviewed (Suslick *et al.*, 1999b). Other studies have turned to the synthesis of carbides, sulfides, and nitrides using the same basic technique. This method of synthesis most effectively uses as starting materials metal carbonyls, which have a finite vapor pressure. Vapor pressure of the metal component inside a bubble formed by acoustic cavitation is key to the transfer of the very high heat generated during the bubble formation and collapse. This *in situ* heat treatment converts the metal carbonyl vapors to nanostructured materials, which are especially active and selective as catalysts (Suslick *et al.*, 1996). The most recent advances have been in the sonication of metal carbonyls in the presence of reactive substrates, such as the formation of nanostructured MoS_2 by sonicating $Mo(CO)_6$ in the presence of elemental sulfur (Mdleleni *et al.*, 1998). Compared to the materials prepared by conventional synthesis, the stoichiometric MoS_2 prepared in this way demonstrated a much higher concentration of edge and defect crystallographic sites. The TEM analysis showed that the sonication-prepared MoS_2 consisted of porous aggregates of clusters of spherical particles 15 nm in diameter, whereas the conventional MoS_2 were large crystallites of well-formed layers. The catalytic evaluation of these materials for the hydrodesulfurization of thiophene demonstrated turnover numbers that were 3 to 5 times greater than those of the conventional catalyst over the range of temperatures studied from 325 to 375°C. Molybdenum carbide, Mo_2C, could be prepared as nanometer grains in surface areas of 163 m^2/g by the sonication of $Mo(CO)_6$ in hexadecane. This catalyst was examined for the hydrodenitrogenation of indole and compared to classically prepared Mo_2C

catalysts as well as Mo_2N (Li *et al.,* 1999a), and the sonication-prepared Mo_2C was slightly more active at low temperatures and less active than the conventional catalysts at high temperatures.

The sonochemical synthesis of Mo_2C was also carried out (Dhas and Gedanken, 1997b) starting with $Mo(CO)_6$ in decane in an inert atmosphere in the presence of activated silica. The resulting Mo_2C crystallites were 5 to 10 nm and coated the silica nonuniformly. A more surprising result (Dhas and Gedanken, 1997a) is the synthesis of unsupported $Mo_2O_5 \cdot 2H_2O$ (Mo^{5+}) when $Mo(CO)_6$ was sonicated in decalin under an air atmosphere. When the same reaction was carried out in the presence of activated silica (Dhas and Gedanken, 1997a, 1997b), spongy platelet nanoparticles were formed in approximately 20-nm grain sizes. The coverage by the Mo^{5+} species was uniform over the surface and tightly bound by Mo–O–Si bonds. Two other interesting results from Gedanken and co-workers (Dhas *et al.,* 1997b) was the observation that nanostructured Cr_2O_3 and Mn_2O_3 could be obtained by the sonochemical-induced reduction of the dichromate and permanganate ions in aqueous media at ambient temperatures. The materials were shown to be amorphous after the sonication; however, calcination of the Cr oxide at 900 K and manganese oxide to 600 K resulted in fine grains of the reduced oxides. The ultrasound-induced synthesis of a novel Pd/C nanostructured composition (Dhas *et al.,* 1997a) resulted when an organometallic Pd precursor was irradiated in argon in mesitylene. The structure of the composition was shown to be a carbon-encapsulated Pd core. The Pd core was 4 to 7 nm, and it was suggested that the high stability of the particles was due to the carbon overlayer.

An extensive study (Emerson *et al.,* 1998) was carried out in our laboratories on the use of high-powered ultrasound to produce nanostructured grains of metal oxides and supported metal oxides. This study examined the effects of different levels of sonication power levels, frequencies, and sonicator reactor designs to determine whether catalysts could be synthesized in controlled grain sizes. All the syntheses included the precipitation step within the high-intensity zone of an ultrasound horn, where the cavitating bubble dynamics are most concentrated. Syntheses of a wide range of nanostructured materials, such as titania, ceria, chromia, nickel oxide, cobalt oxide, alumina, zirconia, bismuth molybdates, iron oxide, and Pt/zirconia, Pt/titania, and Pt/Au alloys on titania, were performed. In all cases, the acoustic cavitational results were compared with results from classical synthesis. The comparison showed that the metal oxide grain sizes could be reduced by 2–3 nm and that, by further calcination of these materials to different levels, catalysts of systematically changing grain sizes in the region 1–20 nm could be achieved. Furthermore, by combining the cavitation experiment with some modification in the chemistry of the precipitation, variable metal oxide grain sizes could be also realized. The study showed that the synthesis of an alloy Pt/Au catalyst over a wide range of metal compositions supported on titania could be realized as pure alloys. A direct comparison to the same range of alloy compositions resulting from the same precipitation without ultrasound power produced a significantly lower degree of alloy formation.

IV. Engineered Synthesis of Nanostructured Catalysts

Recent studies as well as research over the past 20 years in our laboratories have concentrated on the application of chemical engineering reactor design and fluid-flow technology to the synthesis of nanostructured materials as advanced catalysts. The objectives in all these synthetic studies have been to (1) obtain polycrystalline pure-phase materials, (2) synthesize the catalyst primary grains as nanostructured materials, (3) systematically vary the grain size over the range of 1 to 40 nm, and (4) discover and develop methods that are general ones not requiring an exhaustive investigation of the chemistry for each catalyst system desired to be synthesized. For all advanced catalyst work where one is attempting to improve selectivity and reactivity via metal ion modifications of a host metal oxide or alloy, phase purity is the most essential property of the synthesized catalyst. Contamination of the ion-modified catalyst by even small amounts of separate phase materials can lead to overreactivity to undesired products or deactivation of the catalyst. The importance of the nanostructured grain size, as mentioned before, is that many catalytic reactions were reported to show greatly accelerated reaction rates for samples where the catalyst primary grain sizes were in the range 1–10 nm. The importance of a catalyst preparative technique that enables one to systematically change the grain size to any value in the range 1–20 nm is that some reactions show a maximum in reactivity in this size range, and the optimum grain size differs depending on the catalytic reaction.

The classical approach to the synthesis of nanostructured catalysts and materials of different morphologies and grain sizes has been one that heavily relies on detailed experimentation, altering the chemistry of synthesis to achieve the desired material. Our view has been that this approach, although successful for the synthesis of a wide range of nanostructured materials, requires too many experiments and is generally not *a priori* predictable. As a consequence, the direction of nanostructured materials synthesis in our laboratories has been to examine the application of various mechanical techniques that have sufficient process-parameter variability to afford systematic changes in morphology and crystallite size. Inherent in all the processes selected for study was the requirement that the method afford high-phase purity of the synthesized polycrystalline samples of nanostructured catalysts. In all the studies reported here, substantial experimentation was devoted to reactor design for each of the selected techniques to ensure that the fluid flow was well regulated so that the conversion to solid catalysts was accomplished under design conditions.

Aerosol catalyst studies (Michalakos *et al.,* 1995; Moser, 1991; Moser and Cnossen, 1992; Moser and Connolly, 1996; Moser *et al.,* 1993, 1994, 1995a, 1996a; Moser and Lennhoff, 1984, 1989) that we have conducted for more than 20 years concentrated on the development of a well-controlled continuous flow reactor, which had capabilities of regulating the contact time between 0.1 and 8 s, the capability of controlling the temperature profile throughout the reactor, and capabilities for fast quenching of products and for introduction of solid-support materials. After many prototypes were tested, an up-flow reactor design (Moser and Cnossen, 1992; Moser and Connolly, 1996; Moser *et al.,* 1994, 1995a; Moser and Lennhoff, 1989) was finalized. More than 3000 catalysts were synthesized and characterized, and the catalytic properties for many were investigated. In all cases studied, nanometer grains of materials could be synthesized by controlling the fluid-flow conditions through the reactor. The reactor was especially effective for the formation of metastable catalyst structures due to the rapid cooling capabilities. A wide range of catalysts whose compositions included 3 or 4 metal ions demonstrated that materials of exceptionally high-phase purities could be synthesized. The advantage of this aerosol process is that the feed streams used solution of the desired metal salts.

The engineering aspects of our studies on the use of acoustic cavitation to generate nanostructured catalysts centered on the design of a reactor (Emerson *et al.,* 1998) that resulted in the most effective contact of cavitating bubbles with the *in situ* synthesized catalyst gel particle. The equipment permitted control over the bubble dynamics through regulating both the ultrasound's power and frequency. The equipment design also included mixing facilities, which resulted in the synthesis of multimetallic compositions having a high degree of phase purity after calcination.

Our design of a microwave reactor for examining nanostructured catalyst synthesis involved delivering the metal salt solution to the highest energy region of the microwave radiation, where it immediately interacted with a precipitating agent. No cavitation effects are expected using this method; however, the interaction of microwave radiation on a slurry of gel particles might give rise to *in situ* particle heat-up effects, which could afford *in situ* calcination and limiting grain growth. In most cases studied, synthesis under the influence of microwave power resulted in reduced grain sizes after calcination, as contrasted to parallel classical synthesis. For example, in the synthesis of $Bi_2Mo_3O_{12}$, $Bi_2Mo_2O_9$, and Bi_2MoO_6, a comparison of the grain sizes for the microwave to classical synthesis after 598 K calcination revealed grain sizes (microwave/classical) of 15/22, 23/37, and 36/49, respectively. A microwave synthesis of CuO and calcined at 573 K resulted in 13 nm and classical 19 nm; Co_2O_3 synthesis resulted in 13 nm microwave vs 17 nm classical after 573 K calcination. Several other syntheses were carried out; however, the inability to achieve significant reduction in grain sizes does not suggest that this method is useful for variable-grain, nanostructured synthesis at the low power levels of commercial microwave ovens.

A. Hydrodynamic Cavitation

The objective of the research on hydrodynamic cavitation was to use the high Reynolds numbers and controllable throat cavitation numbers to regulate synthesis of both metal oxides and supported metals and metal oxides. From this discussion, it is clear that for catalytic applications, it is important to control phase purity to synthesize homogeneous solid solutions, primary grain (crystallite) size, systematically variable within the range 1–20 nm, and particle (agglomerate) size control and to adjust other properties such as crystallographic strain. In general, it is desirable to obtain the highest degree of mixing of all metal-containing components during a synthesis. This is also important for dual-phase, bifunctional catalysts. Our experimental results and fluid dynamic modeling studies clearly show that controllable synthesis is observed only when the equipment used has the capability of varying the bubble dynamics over a wide range. The technique essentially does catalyst synthesis within the high-shear and high-temperature zone of cavitating bubbles generated by passing a fluid containing the precipitated catalyst precursors through a set of orifices at high velocities and high head pressures. Our general synthetic strategy is illustrated in Fig. 3; this approach applies not only to our studies on hydrodynamic cavitation but also to our studies on acoustic cavitation mediated synthesis.

The schematic in Fig. 3 shows that the synthesis starts with one solution or more containing the metal salts, such as nitrates, acetates, alkoxides, and chlorides. If the solutions are not compatible, the metal ions must be fed using multiple-feed streams. This stream is metered into a second stream containing a precipitating agent such as sodium hydroxide, ammonium hydroxide, or ammonium or sodium carbonate. Care is taken to ensure that the individual metal salts totally and immediately precipitate upon contacting

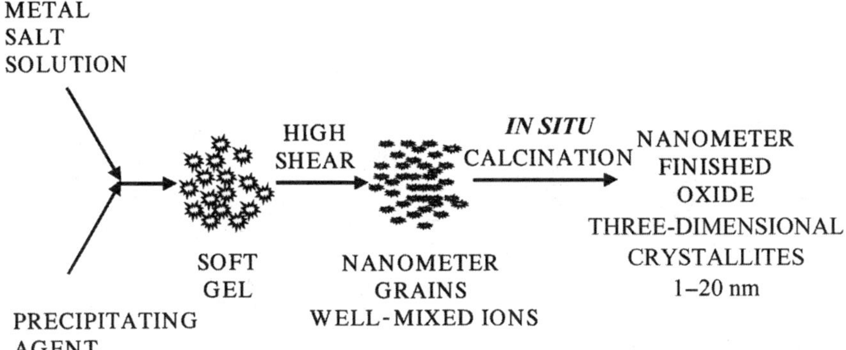

FIG. 3. Schematic of a typical hydrodynamic cavitation synthesis of metal oxides starting with a metal salt solution, which is precipitated and processed in a cavitating stream of bubbles.

the precipitating agent. This results in the formation of a soft gel contained as a slurry within a fast-flowing stream of the precipitating agent. This gel is immediately drawn into the high-pressure zone of the cavitation equipment, where the hydrostatic pressure is regulated between 0.1 and 178 MPa. The stream then passes through an orifice at near-supersonic velocities, which results in the formation of cavitating bubbles. It is the multiple collapse–reformation–collapse of these bubbles in the vicinity of the suspended soft gel that causes the high shear environment and local shock wave heating. This results in both efficient mixing of the individual metal ions and *in situ* calcination to decompose the metal salts to the oxide or hydroxide. The high shear environment also leads to a fragmentation of nanostructured grains of the materials that are agglomerated into larger clusters. The result of this method of synthesis is the formation of nanostructured grains of materials that are normally homogeneous solid solutions in the form of their oxides unless the hydroxides are more stable in aqueous media. Control of the bubble dynamics in the cavitation zone is key to the control of the grain size of the catalysts as well as their agglomerate size.

The equipment that we are currently using for hydrodynamic synthesis was invented by Kozyuk (1996, 1998, 1999a–d, 2000a,b) and manufactured by Five Star Technologies, Cleveland, Ohio. The advantage of this equipment is that it incorporates the process variability necessary to regulate the bubble dynamics required for the synthesis objective. It does this by enabling synthesis to be conducted at very high Reynolds numbers and at variable pressures that are required for adjusting the throat cavitation number and recovery pressure as well as bubble size and shock wave energies in the cavitation zone. This is done by a simple change of the combination of

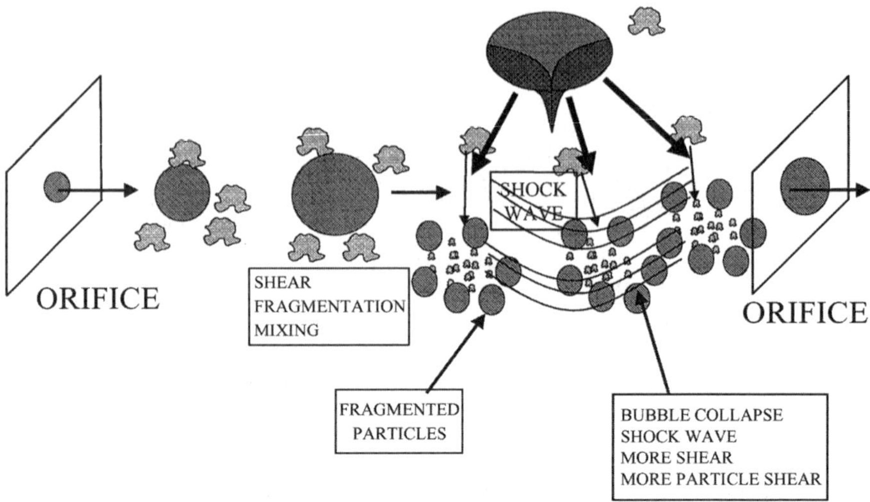

FIG. 4. Schematic of cavitation bubbles interacting with a slurry of precipitated gel particles. The configuration shows typical two orifice processing as afforded in the CaviPro 300 processor. Cavitating bubbles initially form, expand in the recovery zone, and collapse with the formation of a microjet and shock wave.

orifice sizes and geometries located on both the upstream and the downstream side of the cavitation zone. The fluid dynamics problems associated with this method of synthesis are illustrated in Fig. 4. The figure shows the gel slurry entering the first orifice on the left side of the figure. This orifice size, as well as the second orifice on the far right side of the figure, is regulated from 0.050 to 0.50 mm. The hydrostatic pressure of the slurry stream is regulated between 0.1 and 178 MPa, and the fluid flow and Reynolds number are controlled by setting the head pressure and the orifice size utilized in the experiment. When a fast-flowing slurry stream passes through the first orifice, a cavitational bubble is formed due to the Bernoulli effect. Because the second orifice size is usually larger than the first orifice, the pressure in the recovery zone or cavitational zone is less than the head pressure. This pressure difference causes the bubble initially to begin to expand. According to fast photographs taken of bubbles at much lower pressures, the bubbles then begin to turn inward, forming a pointed microjet, and implode with the formation of many smaller bubbles. These bubbles then shrink and rebound; during the rebound process they emit a shock wave (Young, 1989), which gives rise to local heating to within a few bubble radii. These bubbles then implode, and the process is repeated for many cycles. Shear is caused by the bubble formation–collapse–reformation dynamics, and the *in situ* calcination of the solids in the slurry is caused by the shock wave. Because the metal components are not contained within the bubble-gas phase, little calcination occurs due to the high temperature generated within the bubble. The temperature within the bubble resulting from acoustic cavitation has been reported to be within the range of 2000 to 5000 K (McNamara *et al.,* 1999; Suslick, 1993; Suslick *et al.,* 1999a), which is quite effective for decomposing volatile metal carbonyls that have some vapor pressure within the bubble (Hyeon *et al.,* 1996).

The relationship of the double-orifice cavitation generator to the model shown in Fig. 4 is that alteration in the head pressure, fluid flow, and orifice sizes control the bubble dynamics within the recovery zone. Our conceptual model suggests that the bubbles need to be larger than the gel particles, the bubbles need to be under the highest pressure possible, and they must form and collapse as rapidly as possible. These circumstances are optimum for the generation of shear and shock wave calcination of the gel solids. The computational challenge is to calculate the dynamics of the bubble in a flowing system at high pressure where the initial bubble size cannot be estimated. Our fluid dynamic computations on this type of two-orifice system based on relative bubble-size changes show that our device results in the correct bubble-size regime of about a micrometer. However, the results suggest contravening effects. As the pressure within the recovery zone (regime between the first and second orifice) increases, the intensity of the pressure within the bubble increases, and the intensity of the shock wave increases. This gives more intense *in situ* gel particle heating. However, as the pressure in the zone increases, the bubble size decreases. Unfortunately, the fluid dynamic computations on this type of system are too complex currently to enable precise computations to optimize bubble energetics and dynamics. However, based on the relative bubble size computation, we conclude that synthesis studies, where properties such as grain size, agglomerate size, internal strain and phase purity are varied as a function of the orifice parameter set utilized or head pressure applied, will likely show maxima or minima in plots of the experimental results such as grain size vs head pressure. However, at this point the important aspect of the two-orifice system is the capability of the equipment to experimentally vary the dynamics of bubble collapse, and the synthesis studies reported in the next section indicate the important flow regimes.

B. Experimental

Two cavitational devices manufactured by Five Star Technologies were used for synthesis and were the models CaviMax (a single-orifice processor, operating at a constant flow rate and variable pressures) and the CaviPro 300

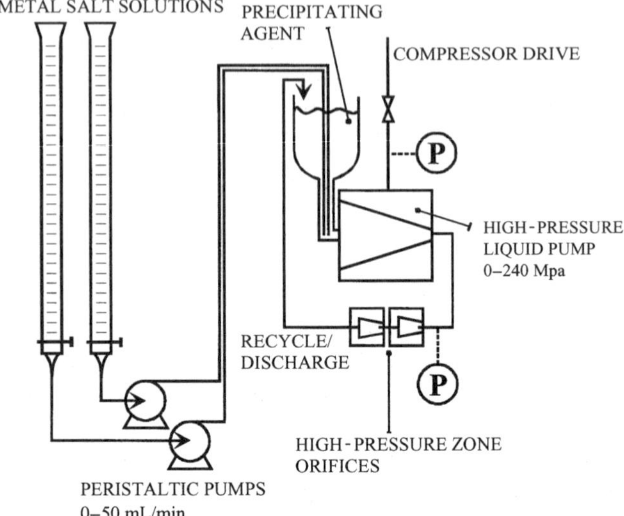

FIG. 5. Diagram of a typical flow configuration for synthesis in the CaviPro 300 processor equipped with a double-orifice cavitation chamber. Metal salt solutions are metered into a precipitating agent just before the inlet to the high pressure pump. The pressure of the solution is then raised to between 1 and 200 MPa and passed through a set of two orifices that generate cavitating bubbles. The effluent is shown to recycle back to the reservoir containing the precipitating agent.

(a double-orifice processor, having variable-pressure-adjustment capabilities). A schematic drawing of the experimental setup is given in Fig. 5.

At the start of a typical experiment, the processor was turned on to begin recirculation of a precipitating agent, which consisted of water or isopropanol containing a dissolved basic salt. Once the precipitating agent was flowing through the system, one or more metal salt solutions were metered into the cavitation device using precision peristaltic pumps, resulting in the immediate formation of a precipitated gel of the metal oxide components. The gel slurry flowed into the processor at the inlet to the high-pressure section and was recirculated during the addition of the feed streams, followed by an additional recirculation period, usually of 10 min. The temperature of the recirculating material was regulated by cooling the gel slurry with an ice bath, in the case of the CaviMax, or through the use of an in-line heat exchanger, in the case of the CaviPro 300. At the end of a synthesis, the product slurry was collected and pressure-filtered under 0.67 MPa of nitrogen on 142-mm-diameter, 0.2-μm nylon filter paper. After the initial pressure filtration, the filter cake was washed and refiltered. The washed solid was then dried overnight (at least 12 h) at 373 K. The dried materials were typically calcined to higher temperatures in air to verify the identity of the product and to make particle size and strain estimates via X-ray diffraction.

X-ray diffraction experiments (XRD) were done with a Rigaku Geigerflex diffractometer (Bragg–Brentano geometry, reflection mode; graphite-secondary monochromator, Cu$K\alpha$ radiation, $\lambda = 154.16$ pm). The program Powder Cell was used to deconvolute phase mixtures and to simulate X-ray patterns. An Amray AMR 1610 scanning electron microscope (SEM), operating at an acceleration potential of 15 kV and an emission current of 65 μA, was used.

The family of $La_{0.6}Sr_{0.4}FeO_3$ perovskites was synthesized using the indicated stoichiometric quantities of the corresponding metal nitrate salts dissolved in water and precipitated with an aqueous solution of Na_2CO_3. A classical precipitation was done in a similar manner using a flask equipped with a mechanical stirrer.

The titania syntheses were carried out using a 0.5 mol/liter titanium tetrabutoxide solution in isopropanol, which was fed into 250 ml of water. After filtration, the samples were dried and calcined at 673 K for 4 h. The PZT samples were prepared using the indicated stoichiometric amounts of titanium tetrabutoxide and zirconium tetrabutoxide in isopropanol, which were cofed with an aqueous lead acetate solution and precipitated using an aqueous $(NH_4)_2CO_3$ solution directly in the inlet of the high-pressure pump. The addition of the solutions was completed after 20 min. The reaction mixture was then processed for an additional 10 min, giving an overall reaction time of 30 min.

C. Characterization of Reynolds and Throat Cavitation Numbers

In order to apply fluid dynamics to understand its effect on synthesis and to use this information to control the synthesis to specific chemical and morphological properties, an extensive study was undertaken to characterize the entire range of Reynolds numbers and throat cavitation numbers affordable by both of the synthesis processors. This investigation used the CaviPro 300 and the CaviMax processors manufactured by Five Star Technologies. The CaviPro 300 is a dual-orifice processor with an operating hydrostatic pressure between 1 and 200 MPa and orifices that can be readily interchanged over the range of 0.050 to 0.50 mm. Naturally, the maximum flow rate of fluid through the instrument and maximum pressure attainable depend on the size of the orifices used in the synthesis experiments. However, the bubble dynamics can be further controlled by establishing the maximum pressure for a given set of orifices and mechanically adjusting the pressure to any lower value. The

CaviMax is a constant, high-flow-rate, single-orifice instrument with a maximum pressure of 8 MPa and orifices that can be interchanged in the range of 1.25 to 2.90 mm. The bubble dynamics can be further controlled through the use of a flow-throttling valve located just after the cavitation recovery zone. Other high-volume processors are available but were not used in this study.

To characterize the fluid flow in the CaviPro 300, a constant second-orifice size of 0.356 mm was used while the first orifice was varied from 0.127 to 0.305 mm. For a given orifice set, the Reynolds and throat cavitation numbers were computed by measuring the flow resulting from adjusting the pressures just before the first orifice from a minimum value to the maximum pressure resulting from that particular orifice set. A wide variety of first orifices were studied, and to illustrate the range, the orifice set 0.127 (first) and 0.356 (second) resulted in Reynolds numbers ranging from 15,000 to 55,000 and throat cavitation numbers ranging from 1.6 to 5.6. Use of the larger first orifice of 0.305 and 0.356 in the downstream (second) position resulted in Reynolds numbers of 22,000 to 55,000 and cavitation numbers of 2.5 to 5.5. For each orifice set, both the Reynolds number and cavitation numbers were precisely adjusted over the indicated range by a simple adjustment of pressure.

Cavitation occurs when bubbles containing either vapor or gas are formed by reduction in the local pressure at constant temperature (Brennan, 1995; Knapp *et al.,* 1979; Young, 1989) such as the rapid movement of the fluid past an impeller blade, through a pump, or in this case through a restriction (orifice) at near supersonic velocities. Quantifying the cavitation number is debatable, but it can be derived from the Bernoulli equation. Lush (Young, 1989) uses a throat cavitation number (σ_{T}) defined as

$$\sigma_T = \frac{p - p_V}{\frac{1}{2}\rho u^2},$$

where p is the total upstream pressure, μ is the liquid velocity in the throat, p_{V} is the vapor pressure of the liquid, and ρ is the liquid density.

Although the onset of continuous cavitation occurs at cavitation numbers σ_T near 0.6, bursts of cavitation occur at 0.3, which begin to cause damage to surfaces. As the number increases, cavitation still may exist, but the bubble dynamics—i.e., size, internal pressure, and periodicity of formation–collapse–reformation—will be different.

Experiments in the CaviMax to characterize the fluid flow varied the orifice size in the single-orifice processor from 1.75 to 6 mm (orifice removed). As mentioned before, the instrument operates at constant flow; thus, the pressure of the measurement decreases as the applied orifice size increases. The flow measurements using this processor resulted in the ability to vary Reynolds numbers between 60,000 and up to 175,000, whereas the range of the available throat cavitation number was 0.85 to 1.7.

In subsequent metal oxide synthesis experiments, the relationship between cavitation numbers and Reynolds numbers to synthesis results, such as crystallographic strain, primary grain size, agglomerate size, or phase purity, was examined to develop an understanding of the effects of different bubble dynamics on crystal properties.

D. Synthesis of Metal Oxide Catalysts and Supported Metals by Hydrodynamic Cavitation

Using the synthetic strategy described previously and depicted in Fig. 3, hydrodynamic cavitation catalyst synthesis experiments were carried out using the continuous-flow system shown in Fig. 5. Two burettes are shown at the left of the figure. Normally only one burette is needed because many syntheses may be carried out by dissolving all the required metal salts in a single solution. However, when salt solutions are incompatible—e.g., one or more components precipitate when mixed—then two or more burettes are required, and in this case, the two solutions containing the different salts are metered into the system at a rate required to achieve the desired stoichiometry. The salt solution is then contacted with a precipitating agent such as aqueous solutions of ammonium carbonate, ammonium hydroxide, metal hydroxides or even a solvent in which all the metal ion components are insoluble. At this point the cavitation processor is in full operation, and the precipitating agent is circulating from the reservoir through the cavitation zone and back to the reservoir. The metal salt solution contacts the precipitating agent at the very bottom of the reservoir; all components immediately precipitate, and the slurry is drawn directly into the high-pressure zone, where the pressure is elevated to the desired head pressure. The set head pressure and orifice set selection determine the Reynolds number and throat cavitation number for the synthesis. The effluent from the cavitation chamber is normally recycled back to the reservoir, but this is done only to achieve the maximum possible effect. Several systems examined showed no difference between a one-pass-through experiment and a recycle one. For commercial operations requiring high throughput, the single-pass mode would be used in most cases. In the recycle mode, the metal salt addition is usually done over 20 min, and another 10 min is allowed for total recycle. After synthesis, the solids are either isolated by high-rpm centrifuging or pressure filtration in 1-liter stainless steel filters equipped with a 0.2-μm nylon filter. All solids were dried at 298 K for at least 12 h. Depending on the system, X-ray diffraction, transmission electron microscopy, and X-ray photoelectron spectroscopy were measured on either the dried samples or samples that had been calcined in air to different temperatures.

E. Estimation of the *in Situ* Calcination Temperature in MoO_3 Synthesis

An important aspect of materials synthesis by hydrodynamic cavitation is the degree of *in situ* calcination of the sample while it is flowing in a liquid medium in a high-shear environment. To estimate this heating, the synthesis of MoO_3 was investigated using several cavitational conditions and both processors. These results were compared to the synthesis by simply heating the starting material, ammonium heptamolybdate, $(NH_4)_6MoO_7 \cdot 7H_2O$, to eight different temperatures from 373 up to 598 K in a calcining oven in air for 4 h. When the ammonium molybdate is heated to different temperature levels, several different intermediates are generated, whose phases are well characterized in the literature. To estimate the average bulk temperature to which the solids are heated for the extremely short residence times in the flowing stream through the cavitation generator, the cavitation processed materials were air-dried at 373 K and their phase composition was compared with those of the oven-prepared materials at the different temperatures. To approximate the oven preparation in the cavitation processor, the ammonium molybdate was dissolved in a small amount of water and fed into a large volume of isopropanol, which contained a low concentration of ammonium hydroxide (0.03 *M*) to ensure complete precipitation. The precipitation technique was used rather than simply passing a slurry of the insoluble crystals through thc processor to better simulate the grain size of processed insoluble material and the bubble size defined by the process conditions. A variety of cavitation conditions were studied in both the CaviPro 300 and CaviMax processors. After the cavitation synthesis, the solid materials were filtered and air-dried at 373 K, the XRD was measured, and a complete phase analysis was made on each similarly treated sample. A complete phase analysis was performed on all the oven-calcined samples at the different temperatures. The latter analysis provided information on the composition of the different species resulting from a 4-h calcination for comparison to the cavitation processed samples, which had been dried at 373 K. In addition, kinetic data were obtained for the decomposition of one of the stable intermediates because it appeared in the XRD from samples at several temperatures. These reactivity data at different temperatures were used to qualitatively evaluate the activation energy for the decomposition of the oven-calcined samples. This activation energy was then applied in the Arrhenius equation, along with the residence time of the fluid element inside the cavitation section (recovery zone) to determine short-lived *in situ* heat-up temperature from the shock wave. The effective bulk temperature of *in situ* heating of the particles in the cavitation experiments was obtained by comparing the phase observed after the 373 K drying to the temperature required to give

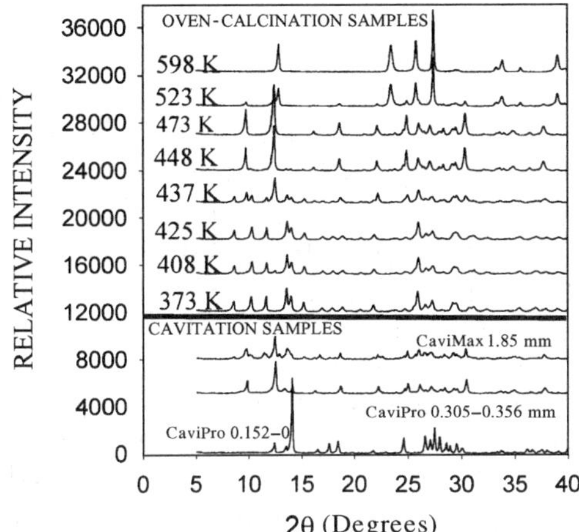

FIG. 6. XRD analysis of oven-calcined ammonium molybdate for 4 h at progressively higher temperatures are shown in the top half of the figure. The XRD of 373 K–dried products from the treatment of ammonium molybdate in an inert solvent in the CaviPro 300 and CaviMax processors are shown in the bottom half of the figure.

the same phase in the oven-calcined series. Three cavitation-prepared samples were compared to the eight oven-calcined samples in Fig. 6 by XRD. The top eight diffraction patterns resulted from the oven calcinations to the temperatures indicated on the graph. The lower three patterns resulted from processing ammonium molybdate using the CaviPro equipped with a 0.152-mm orifice in the first (upstream) position and no orifice in the second position. Also, the XRD of the cavitation sample processed with the orifice set 0.305–0.356 mm is displayed along with a CaviMax experiment that used the 1.85-mm orifice. After cavitational processing, all the samples were dried at 373 K and analyzed by XRD.

The sample obtained from the lowest temperature oven calcination at 373 K was identified as having a main phase of $(NH_4)_8Mo_{10}O_{34}$, N/Mo = 0.8 (ICDD, 1997b). In the oven-calcined series, the 473 K sample was identified as having a main phase of $(MoO_3)_4(NH_3)_2(H_2O)$, N/Mo = 0.5, according to the ICDD data base (ICDD, 1997c). The cavitation processed sample using the single 0.152-mm orifice and dried at 373 K was identified as triclinic ammonium molybdenum oxide $(NH_4)_4Mo_8O_{26}$, N/Mo = 0.5 (ICDD, 1997a). The composition of the oven-calcined sample at 473 K for 4 h had the same composition as the CaviPro sample processed for short residence times; namely, the N/Mo ratios for both were 0.5 and the O/Mo ratios were 3.25. Thus, their compositions are the same and their crystallographic structures are very similar, with the triclinic phase resulting from the cavitation treatment likely being of slightly higher energy. From these data, we conclude that the effective average bulk temperature resulting from the shock wave heating was approximately 473 K. For the CaviPro sample processed using the dual orifice set 0.305–0.356 mm, the estimated average bulk temperature heating was around 448 K, and processing in the CaviMax using the single 1.85-mm orifice resulted in an effective heating to 438 K.

Using the approximate phase composition data obtained from the 425, 437, 448, and 473 K oven-calcined samples in the middle section of Fig. 6 and assuming a first-order solid-state decomposition of ammonium molybdate, an activation energy of around 92 kJ/mol was computed. In the CaviPro synthesis, the number of recycles was 4.16 passes of the 600-ml solution flowing at a rate of 125 ml/min through the cavitation chamber. Because the volume between the two orifices was 0.20 ml (in the recovery zone), the total contact time for the ammonium molybdate solution was 0.385 s in the recovery zone. Using the oven-calcination experiment at 473 K, where the conversion to the $(MoO_3)_4(NH_3)_2(H_2O)$ intermediate phase was at least 95%, assuming first-order kinetics and a total reaction time of 4 h results in a first-order rate constant of 1.599×10^{-4} s^{-1}. Because the degree of conversion to an intermediate of the same stoichiometry, $(NH_4)_4Mo_8O_{26}$, was obtained from the cavitation experiment, a first-order rate constant of 0.385 s^{-1} was calculated, assuming that all the shock wave heating occurred within the 0.20-ml reaction zone between the orifices and that the reaction maintains the same order—i.e., first order, as in the oven-heated samples of ammonium molybdate. Applying these rate data to the Arrhenius equation and making the calculation assuming that the preexponential factor and activation energies are the same for both reactions, the effective *in situ* shock wave average heat-up temperature that the gel particles experience is around 700 K. Naturally, the temperature of the shock wave would be much higher than this number, but average effective heat transferred corresponds to heating the particle to 700 K. This estimate is regarded as conservative because the time constant for bubble collapse–reformation–collapse is likely much less than the residence time computed using the volume of the cavity. Thus, a more realistic residence time to use in the first-order kinetic treatment would be the time constant for the shock wave interacting with each solution element. As a limit in calculating the minimum *in situ* heat-up temperature, the velocity of sound through liquid water at 298 K was used in combination with the solution volume and the 4.16 recycle passes to compute the first-order rate constant. Calculation of the reaction rate based on these data and applying the rate to the Arrhenius equation resulted in an average *in situ* shock wave transient heat-up temperature of 950 K. These values for the *in situ* heat-up temperature should not be compared to the values of 2000 to 5000 K reported by Suslick and co-workers for the heat-up temperatures within the bubble resulting from acoustic cavitation (McNamara *et al.,* 1999; Suslick, 1993;

Suslick, *et al.,* 1999a). Cavitational heating results from heating within the bubble and shock wave heating. Heating within the bubble is regarded as ineffective for heating components located outside the bubble, whereas shock wave heating is responsible for heating components within a few radii of the bubble. Thus, we conclude that the effective shock wave *in situ* calcination temperature lasting for very short heating times is in the range of 700 to 950 K, and this will vary, depending on the metal oxide being synthesized due to the differences in heat-transfer coefficients for the metal oxide gel component.

A second method for acquiring evidence for shock wave heating was gained from passing a screened sample of calcium fluoride, CaF_2, through a CaviMax processor in water and recirculating for 30 min. The scanning electron micrograph of the untreated CaF_2 is shown in Fig. 7, and the CaviMax-processed sample is shown in Fig. 8. The melting point of CaF_2 is 1360°C. A comparison of the SEM of the cavitation-processed sample and unprocessed sample shows that the processed sample in Fig. 8 contained mainly CaF_2 crystals, which were rounded off, and many particles were attached to one another by melted necks. This suggests that the particles were heated to a sufficiently high temperature so that a major fraction of the particle's

FIG. 7. A sample of calcium fluoride (CaF_2) that had been ground and screened to approximately 2 μm.

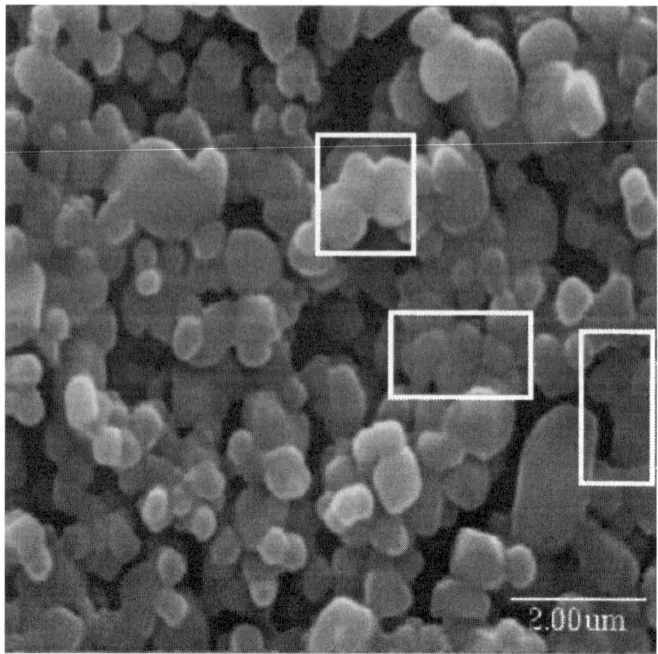

FIG. 8. A sample of the CaF_2 that had been screened to 2 μm (seen in the SEM in Fig. 7) which was processed in the CaviMax processor with recirculation as a slurry in water for 30 min. The highlighted areas indicate particles having melted necks.

surface had melted or at least softened. One can expect that CaF_2 would soften at the Tamman temperature of about 50% of its normal melting point. Thus, a shock wave heat-up temperature to about 600°C is consistent with the data for ammonium molybdate decomposition. These results are similar to those observed (Suslick, 1993) for shock wave heating in ultrasound experiments, where similar melting was observed after high-melting solids were processed in an acoustic cavitational stream.

F. Hydrodynamic Cavitation Synthesis of Nanostructured Catalysts in High-Phase Purities and Varying Grain Sizes

One of the advantages of cavitational synthesis, providing the synthesis equipment has the capability to adjust the bubble dynamics, is the capability to do synthesis under the conditions of both high shear and *in situ* calcination. If the equipment can be adjusted to operate in slightly different fluid dynamic regimes, it should be possible to obtain both high-phase purity catalysts and to systematically adjust the primary grain size of the crystallites

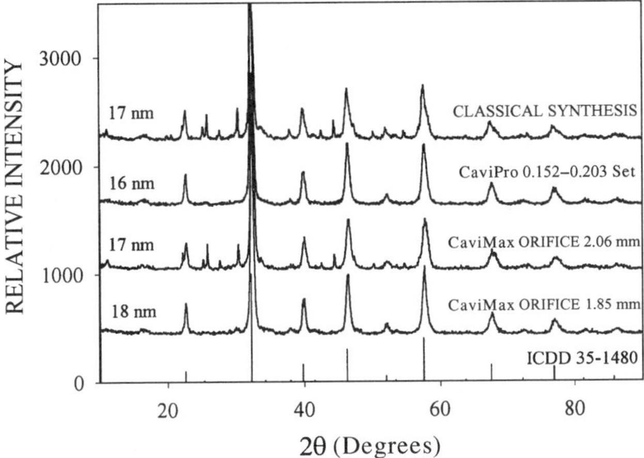

FIG. 9. XRD of the CaviMax and CaviPro 300 synthesis of $La_{0.6}Sr_{0.4}FeO_3$ compared to the classical synthesis.

in the important region of 1 to 20 nm. Syntheses of several complex metal oxide systems were studied in order to determine whether hydrodynamic cavitation had the capabilities to meet the previously mentioned synthetic objectives.

The synthesis of β-$Bi_2Mo_2O_{12}$ (β-bismuth molybdate) and a multimetallic perovskite, $La_{0.6}Sr_{0.4}FeO_3$, were examined to determine whether the cavitational synthesis afforded both superior phase purities and variable grain sizes, as contrasted with classical methods of coprecipitation synthesis. The perovskite synthesis was extensively examined in the CaviPro 300 and the CaviMax using a variety of different orifice sets to intentionally vary the bubble dynamics. In all cases, the products were isolated by filtration and calcined to different temperatures for 12 h, after which their grain sizes and phase purities were examined by XRD. The XRD patterns in Fig. 9 resulted from catalysts synthesized in the CaviMax processor using the 1.85-mm orifice (lowest curve) and the 2.06-mm orifice (second curve from the bottom). The CaviPro 300 processor was used with several different orifice sets, and the XRD resulting from the 0.152-mm (first orifice) and 0.203-mm (second orifice) set is illustrated in the third curve from the bottom. For comparison purposes, the XRD of a coprecipitation carried out under conditions of high-speed mechanical mixing with the identical chemistry and precipitation technique is illustrated in the top diffraction pattern in the figure. All materials in the figure were calcined to 873 K for 12 h after synthesis. The XRD data show that the two CaviMax synthesized materials gave substantially different results; namely, the smallest orifice size, 1.85 mm, resulted in both very high phase impurity and a grain size of 18 nm, whereas the 2.06-mm orifice afforded a material having substantial amounts of impure phases in a grain size of 17 nm. Synthesis using the 2.26- and 2.43-mm orifices in two other CaviMax experiments resulted in the same low degree of phase impurity as observed using the 2.06 orifice, and the grain sizes were 17 and 16 nm, respectively. However, synthesis using the CaviPro 300 processor resulted in high-phase pure materials for all experiments using a constant size for the first orifice while varying the size of the downstream orifice. The orifice sets studied were 0.152–0.203, 0.152–0.254, 0.152–0.305, and 0.152–0.356. The XRD data using the 0.152–0.203 sets are illustrated in the third curve from the bottom of Fig. 9. The XRD data for the classically prepared material are illustrated in the top curve in the figure, and this pattern shows that large amounts of separate phase compounds were observed in the sample. The grain size of the CaviPro 300–prepared materials was 16 nm, whereas the classical was 17, and the grain sizes resulting from all four of the orifice sets studied were between 16 and 18 nm. Thus, although the cavitational synthesis resulted in high phase purities and the classical synthesis did not, the fluid dynamic region studied was not able to systematically vary the grain size of the calcined materials.

The synthesis of β-$Bi_2Mo_2O_{12}$ using both the CaviMax processors afforded results that paralleled those of the perovskite synthesis. The phase purity of the catalysts after calcination at 623 K by cavitation using the 1.85-mm orifice gave phase-pure $Bi_2Mo_2O_{12}$, whereas the parallel classical synthesis contained a modest amount of the alpha phase. Processing using the 2.06-, 2.26-, and 2.41-mm orifices all resulted in a modest amount of impure phase similar to the classical synthesis and grain sizes after 623 K calcination of 22, 28, and 21 nm, respectively. The grain size observed using the 1.85-mm orifice was 11 nm, whereas the classical synthesis resulted in 19 nm.

G. The Introduction of Crystallographic Strain in Catalysts by Hydrodynamic Cavitation

During our studies on hydrodynamic cavitation-mediated catalyst synthesis, we observed the introduction of crystallographic strain in several metal oxide systems. As the size of a crystallite decreases into the range of a few nanometers, both crystallographic strain and defects in metal oxides increase, and it was shown that higher defect concentrations in metal sulfides result in higher HDS catalytic reactivity (Mdleleni *et al.*, 1998). The degree of strain in nanostructured materials synthesized by cavitation was characterized by the Williamson–Hall XRD (Williamson and Hall, 1953) technique. The strain induced during cavitational synthesis was investigated for titania, several piezoelectric (PZT) systems in the family $PbZr_xTi_{1-x}O_3$, and the synthesis of an alumina-stabilized copper-modified zinc oxide. A wide variety of fluid dynamic conditions were investigated to understand the relationship between strain and cavitation. The expectation was that as the primary grain size decreased, the strain would increase. After collection of the data for the titania and PZT, a classical engineering technique, the Buckingham (1914) pi theorem was applied to the systems for predicting flow regimes that result in the highest degree of crystallographic strain.

1. Titania

Because strain in nanostructured titania should affect both the electronic and catalytic properties of the materials, an extensive study of the cavitational synthesis of titania was carried out. The synthesis of titania starting with titanium butoxide, precipitated in either pure water or using stoichiometric to various excess amounts of water in isopropanol, was studied in both the double-orifice processor (CaviPro 300) and the single-orifice processor (CaviMax). The syntheses in the single-orifice processor were carried out in pure water by running a 0.5 *M* solution of the butoxide in isopropanol into a recirculation stream of water. The XRD analysis of these materials after calcination at 673 K showed an unusual degree of induction of crystallographic strain as compared to the materials prepared by classical precipitation, and the amount of strain depended on the fluid dynamic conditions applied in the synthesis. The XRD data for titania synthesis using the single-orifice processor equipped with six different orifice diameters from 1.85 to 1.92 mm are shown in Fig. 10. Included in the figure are the grain sizes observed after a 673 K calcination. Although the Reynolds numbers and throat cavitation numbers were varied over a wide range, the primary grain sizes were nearly constant. However, the strain systematically increased up to 0.34%, which is much higher than the classically prepared material of 0.25% with a grain size

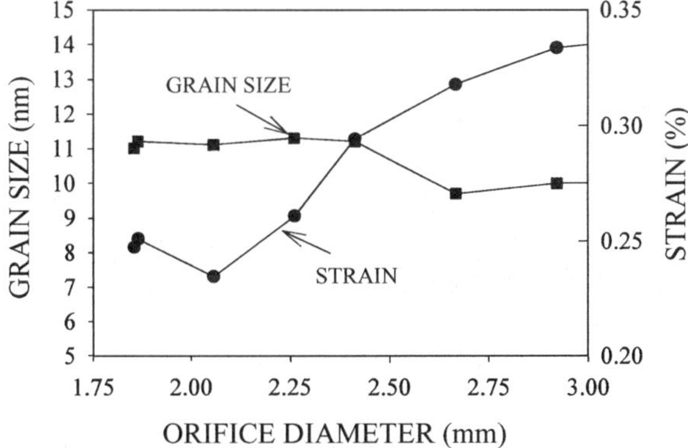

FIG. 10. Grain size and crystallographic strain data for the synthesis of titania using the CaviMax processor as a function of different orifice sizes used in the processor.

of 14 nm compared to the 11-nm grains in the cavitation experiments. Using the smallest orifice shown in Fig. 10 (1.85 mm), several syntheses were carried out where titanium butoxide in isopropanol was reacted with a circulating solution of water in stoichiometric amounts up to 16 times the stoichiometric ratio. These syntheses resulted in slightly lower strain, near 0.25%, whereas the classical preparations afforded strains of 0.20% and grains of 15 nm. Synthesis in the two-orifice processor using a downstream constant orifice size of 0.356 mm while the first orifice was varied in four ranges from 0.178 to 0.305 mm resulted in a substantially higher strain of 0.55% and slightly smaller grain sizes of 10 nm. The grain sizes were constant irrespective of the size of the first orifice, whereas the strain was highest using the smallest first orifice.

It is clear from the preceding results that cavitational synthesis generates substantially more strain in the titania crystallite than classical synthesis and that the high-pressure experiment (dual-orifice processor) introduces more than the low-pressure synthesis. Furthermore, the amount of strain can be systematically varied by changing the processor's operating parameters. To aid in understanding the relationship between the fluid-flow properties of the cavitational synthesis and the degree of strain induced in the titania synthesis, a Buckingham (1914) pi treatment was applied to the titania synthesis data obtained from the single-orifice processor. This dimensional analysis of the pertinent dimensionless variables (Find *et al.,* 2000) resulted in a three-dimensional surface with a correlation coefficient of 0.92, as shown in Fig. 11.

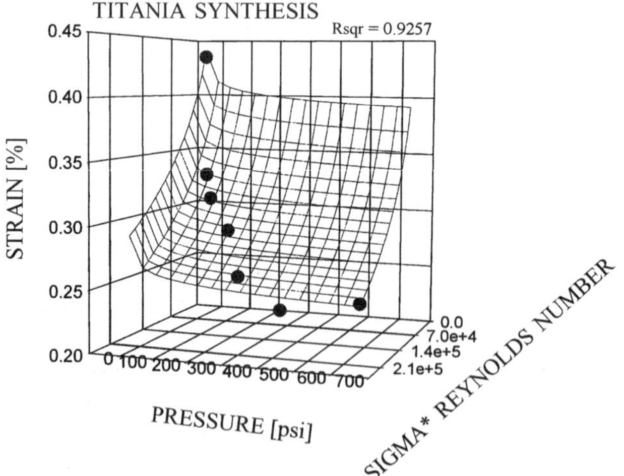

FIG. 11. Three-dimensional surface plot showing the correlation of applied pressure, crystallographic strain, and Reynolds number–throat cavitation number for the synthesis of titania using the CaviMax processor.

After the analysis was completed, the plot was used to predict the processor conditions required to synthesize the point of highest strain in the figure of 0.44%. A power law fit of the data showed that the microstrain was more sensitive to changes in the Reynolds number, but the strain decreased as the Reynolds number increased. Over the regime studied, the microstrain increased as the cavitation number decreased for $\sigma_T = 0.5$ to 1.8.

A series of experiments were carried out to determine whether cavitational processing of a preformed nanostructured titania could further reduce the primary grain size and reduce the agglomerate size. For this experiment a sample of titania previously calcined to 673 K was recirculated through the dual-orifice processor using the 0.254–0.356 orifice set at a constant pressure of 120 MPa in two experiments, one in water and the other in *n*-hexane. Each system was studied over 4 h of processing, the starting titania was sieved to agglomerates of greater than 1000 nm, and the primary grain size was 13 nm. In the case of the water system, after 1 h, 65% of the agglomerated particles had been reduced to below 1000 nm and the distribution was centered at 600 nm. After 3 h, 72% of the agglomerates had been reduced to below 1000 nm, with agglomerates centered at 300 nm. Processing in hexane resulted in 74% agglomerates below 1000 nm after 1 h, with the distribution centered at 300 nm. After 4 h 88% of the particles were below 1000 nm, and the distribution was bimodal (centered at 150 nm and 500 nm). The grain size of the titania remained unchanged from the original 13 nm after any of the processing. An interesting feature of the materials processed in water is that the XRD intensities of the anatase peaks linearly decreased with processing time to 50% of their original intensity after 4 h. Because the grain size did not change and the background (indicating amorphous materials) did not change very much, we conclude that the decline in intensity was due to a lower materials density of the processed solids. After the initial processing time, the microstrain increased linearly with increased processing time, and the amount of rutile in the sample systematically increased from 0 to 10%.

2. *Piezoelectrics*

A wide range of studies in the single- (CaviMax) and double-orifice processors (CaviPro 300) on the piezoelectric (PZT) system in the family $PbZr_xTi_{1-x}O_3$ showed that the processing parameters could greatly enhance the amount of microstrain to values as large as 0.93%. Figure 12 illustrates the range of microstrain that resulted from cavitational processing in both the single-orifice and double-orifice processors. The microstrain was measured on samples calcined in air at 873 K. The figure shows that the high-flow-rate, single-orifice processor resulted in a much higher degree of induced strain than the low-flow, high-pressure single-orifice processor. In the figure, the

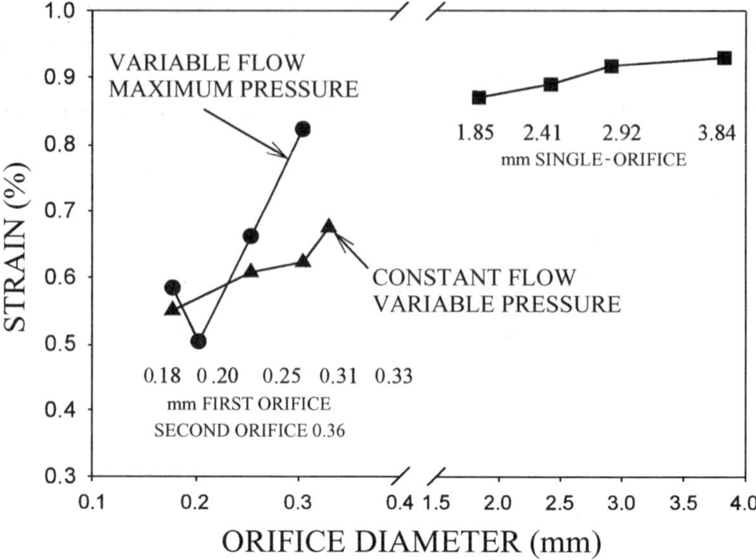

FIG. 12. Strain introduced into PZT crystallites of $PbZr_{0.5}Ti_{0.5}O_3$ by cavitational processing in the double-orifice processor (CaviPro 300) as a function of different sets of orifices, shown on the left side of the figure. Similar data using the single orifice processor (CaviMax) are shown on the right side of the figure.

range of minimum to maximum pressure for the double-orifice, variable-flow, maximum-pressure experiments was 20.6 to 169 MPa. The flow rate in the constant-flow, variable-pressure experiments used constant flow rates of 400 ml/min by adjusting the operating pressure of the processor. Other experiments showed that all the cavitational synthesized PZT compositions were essentially 100% pure phase, whereas the classically prepared materials all contained modest amounts of secondary phases. The grain sizes of the cavitational samples after 873 K air calcination were around 25 nm. Although it has not been reported experimentally in the literature, it would be expected that these high levels of microstrain would affect both the piezoelectric properties and the rates of grain growth.

3. Copper-Modified Zinc Oxide

Several series of methanol synthesis catalysts were prepared by the coprecipitation of the appropriate metal salts in either Na_2CO_3 or $(NH_4)_2CO_3$ using the CaviPro 300 (dual-orifice processor) at both high and low pressures. The experiments were carried out using a 0.152-mm first orifice while varying

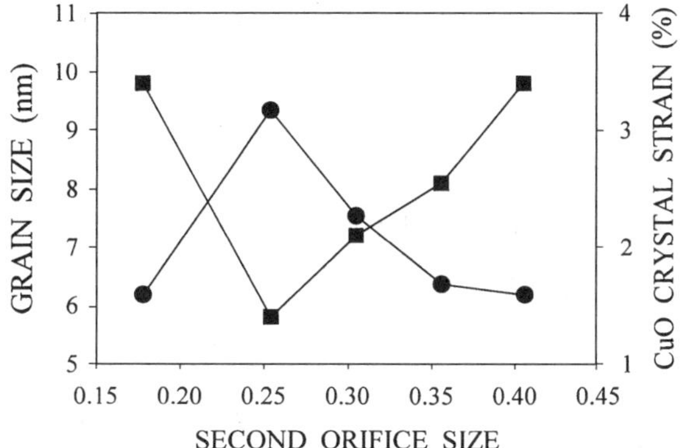

FIG. 13. Grain sizes and crystallographic strain in CuO after 623 K calcination of $Cu_{0.22}Zn_{0.68}Al_{0.10}$ catalysts from a CaviPro 300 synthesis using a 0.152-mm first orifice while varying the second orifice between 0.178 and 0.406 mm. The composition of the catalyst was $Cu_{0.22}Zn_{0.68}Al_{0.10}$.

the second orifice between 0.178 and 0.406 mm at a constant head pressure of 138 MPa. The resulting catalysts had the composition of $Cu_{0.22}Zn_{0.68}Al_{0.10}$ and were calcined at 623 K in air before XRD analysis to determine both the crystallographic strain and the grain size of the CuO component. A similar analysis was performed on the ZnO component, but both the grain sizes and the strain were essentially constant as a function of the orifice set applied in the synthesis. However, both the grain size and the strain of the CuO component varied as the orifice size of the second orifice was varied. Figure 13 illustrates that the cavitational synthesis results in the capability to systematically vary not only the grain size of the CuO component, but also the crystallographic strain. The strain increased as the grain size of the CuO decreased. In view of the report (Doesburg *et al.*, 1987) that methanol synthesis rates using nanometer-sized CuO catalysts varied according to the grain size, this ability to systematically vary the CuO grain size is likely important for the synthesis of catalysts having the appropriate grain size corresponding to the highest catalytic rates of methanol formation.

H. Synthesis under Variable Fluid-Flow Conditions

By a simple mechanical adjustment in the orifice sizes, type of processor, or the head pressure, the cavitation bubble dynamics can be systematically altered. When catalyst synthesis is carried out under different conditions obtained by this type of adjustment, the structures and grain sizes of products vary considerably. This section gives a few examples of the divergent synthesis results for a few of the materials that were synthesized.

1. Synthesis of High-Temperature Stable Oxidation Catalysts

The synthesis of catalysts for high-temperature gas turbine applications was carried out to obtain a 2% w/w Pd on zirconia–alumina. The synthesis used 5 mol% alumina as a stabilizer to prevent grain growth of the zirconia support. In this synthesis, all components were precipitated in a single step using the CaviMax processor equipped with the 1.85-, 2.06-, 2.41-, or 2.92-mm orifice. The products after filtration, drying, and calcination to 1373 K were analyzed by XRD; the diffraction patterns are illustrated in Fig. 14. A classical experiment was carried out using rapid mechanical stirring in the precipitation step and is illustrated in the top diffraction pattern of the figure. It is interesting to note that these conditions resulted in the formation of mainly the monoclinic phase of zirconia in the classical preparation. The cavitational preparation formed mainly this same phase using the smallest orifice (1.85 mm), but as the orifice size increased up to 2.92 mm, the calcined materials were mainly a cubic zirconia containing a small amount of the monoclinic phase. Although the Pd reflection at 40.25° 2θ is weak in all cases, it can be seen from the figure that the reflection grew when the 2.92-mm orifice was used. Quantification of the grain sizes was performed using long analysis times and showed that the 2.41-mm experiment

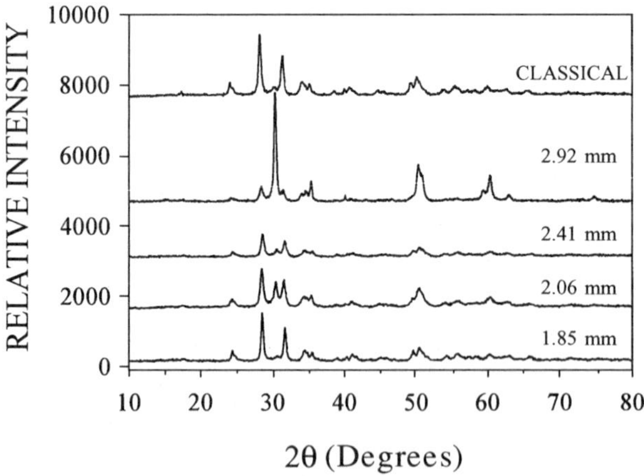

FIG. 14. XRD results for the synthesis of 2% Pd on zirconia–5 mol% alumina after 1373 K calcination using the CaviMax processor equipped with either the 1.85-, 2.06-, 2.41-, or 2.92-mm orifice. The corresponding classical preparation is illustrated at the top of the figure.

resulted in a Pd grain size of 36 nm, whereas the Pd grain size in the 2.92-mm orifice experiment was about 100 nm.

2. Synthesis of Variable Phases under Different Bubble Dynamics

The synthesis of 2% Ag on alumina was studied in both the CaviMax and CaviPro 300 processors under a wide variety of bubble dynamic conditions. Several catalysts were synthesized by a single-step precipitation of both the metal and the alumina support, which passed into the high-pressure pump of the processor. In the CaviPro 300, the orifice sets 0.127 mm/0.356 mm, 0.178 mm/0.356 mm, 0.229 mm/0.356 mm, and 0.279 mm/0.356 mm were studied in recycle experiments. After the products were dried at 298 K overnight, the XRD of the catalysts showed that when the smallest first orifice in the set, 0.127 mm/0.356 mm, was used, the product was nearly pure Böhmite, whereas using progressively larger first-orifice diameters, as in the 0.178 mm/0.356 mm experiment, resulted in mainly Bayerite. The XRD analysis on these materials is shown in Fig. 15. A similar set of experiments using different orifices sizes in the CaviMax single-orifice processor resulted in a similar variability in alumina structure. In all cases, the silver grains were very small, as seen by inspecting the weak reflections near 64° 2θ. The material from the experiment that afforded mainly Bayerite was transformed to η-Al_2O_3 when air-calcined to 673 K and θ-Al_2O_3 when calcined to 1073 K. The material exhibiting mainly Böhmite reflections was transformed to γ-Al_2O_3 upon calcination to 673 K and δ-Al_2O_3 when calcined to 1073 K. Thus, this type of synthesis has the capability of affording nanostructured

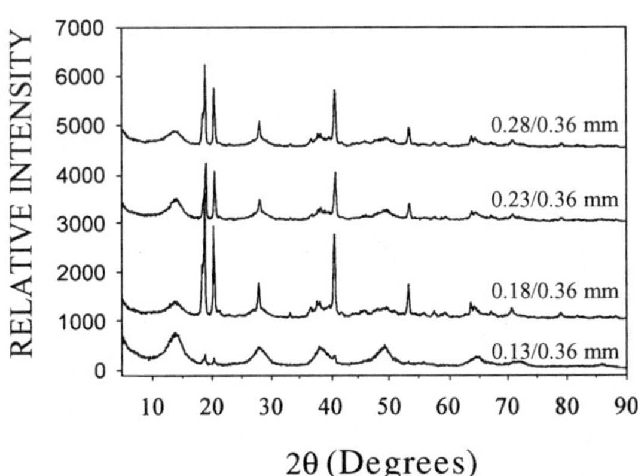

FIG. 15. XRD diffraction patterns for the synthesis of Ag 2% w/w on alumina after drying at 373 K. The catalysts were obtained from processing in the CaviPro 300 using the orifice sets 0.127 mm/0.356 mm, 0.178 mm/0.356 mm, 0.229 mm/0.356 mm, and 0.279 mm/0.356 mm.

grains of Ag supported on nanometer-size grains of various transitional aluminas by conducting the synthesis under different hydrodynamic cavitation conditions.

V. Conclusions

A large number of novel techniques for the synthesis of nanostructured materials have been developed in recent years. Most of the processes rely on either chemical techniques or mediated synthesis by polymers or surfactants to produce nanostructured grains of catalyst that are stable. Few processes result in the general ability to systematically vary the primary crystallite grain sizes in the important range 1–20 nm, which is critical for the discovery of active and selective catalysts. Many processes result in the formation of multimetallic catalysts in high phase purities, which is just as important for catalysis. Hydrodynamic cavitation processing affords the capability to calcine precipitated products during the synthesis step and to systematically introduce crystallographic strain. In some cases, the hydrodynamic cavitation process afforded variable catalyst grain sizes in the range 1–10 nm by a simple mechanical adjustment in the process parameters, but metal oxide synthesis did not generally result in a large variability of grain sizes.

ACKNOWLEDGMENTS

We thank Five Star Technologies of Cleveland, Ohio, for generous financial support of this research. We also acknowledge Soonan Nguyen, Tri Giang, Vinh Voduc, and Hieu Nguyen for their help in some syntheses and XRD analyses.

REFERENCES

Allpress, J. G., and Sanders, J. V., Structure and stability of small clusters of atoms. *Aust. J. Phys.* **23,** 23 (1970).

Baetzold, R. C., Calculated properties of metal aggregates. I. Diatomic molecules. *J. Chem. Phys.* **55,** 4355 (1971).

Baetzold, R. C., Molecular orbital description of catalysis by metal clusters. *J. Catal.* **29,** 129 (1973).

Behrens, S., Dinjus, E., and Unger, E., Protein supported metallic nanostructures. *Nachrichten (A publication of the Forschungszentrum Karlsruhe, Germany),* **31,** 117 (1999).

Behrens, S., Habicht, W., Boukis, N., Dinjus, E., Baum, M., and Unger, E., Private communication from S. Behrens, Institut fur Technische Chemie, Forschungszentrum Karlsruhe, Germany (2000).

Berthet, A., Thomann, A. L., Aires, F. J. C. S., Brun, M., Deranlot, C., Bertolini, J. C., Rozenbaum, J. P., Brault, P., and Andrezza, P., Comparison of bulk Pd/(Bulk SiC) catalysts prepared by atomic beam deposition and plasma sputtering decomposition: Characterization and catalytic properties. *J. Catal.* **190,** 49 (2000).

Birringer, R., Gleiter, H., Klein, H. P., and Marquardt, P., Nanocrystalline materials—An approach to a novel solid structure with gas-like disorder. *Phys. Lett. A* **102,** 365 (1984).

Brennan, C. E., "Cavitation and Bubble Dynamics." Oxford Univ. Press, New York (1995).

Buckingham, E., On physically similar systems: Illustrations of the use of dimensional equations. *Phys. Rev.* **4,** 345 (1914).

Busser, G. W., van Ommen, J. G., and Lercher, J. A., Preparation and characterization of polymer-stabilized rhodium particles, *in* "Advanced Catalysts and Nanostructured Materials, Modern Synthetic Methods" (W. R. Moser, Ed.), p. 213. Academic Press, San Diego (1996).

Cox, D. M., Kaldor, A., Fayet, P., Eberhardt, W., Brickman, R., Sherwood, R., Fu, Z., and Sondericher, D., Effect of cluster size on chemical and electronic properties, *in* "Novel Materials in Heterogeneous Catalysis" (R. T. K. Baker and L. L. Murrell, Eds.), Vol. 437, p. 172. Amer. Chem. Soc. Symp. Ser. Washington, DC (1990).

Crooks, R. M., Zhao, M., Sun, L., Chechik, V., Yeung, L. K., and Lemon, B. I., Dendrimer-encapsulated nanoparticles: Synthesis, characterization, and applications to catalysis, private communication.

de Heer, W. A., The physics of simple metal clusters: Experimental aspects and simple models, *Rev. Modern Phys.* **65**(3), 612 (1993).

Del Vallee, M., Avalos-Borja, M., Cruz, J., and Fuentes, S., Molecularly designed ultrafine/nanostructured materials, *Mater. Res. Soc, Symp. Proc.* **351,** 287 (1994).

Dhas, N. A., and Gedanken, A., Characterization of sonochemically prepared unsupported and silica supported nanostructured pentavalent molybdenum oxide, *J. Phys. Chem. B* **101,** 9495 (1997a).

Dhas, N. A., and Gedanken, A., Sonochemical synthesis of molybdenum oxide—and molybdenum carbide–silica nanocomposites. *Chem. Mater.* **9,** 3144 (1997b).

Dhas, N. A., Cohen, H., and Gedanken, A., *In situ* preparation of amorphous carbon-activated palladium nanoparticles. *J. Phys Chem. B* **101,** 6834 (1997a).

Dhas, N. A., Koltypin, Y., and Gedanken, A., Sonochemical preparation and characterization of ultrafine chromium oxide and manganese oxide powders. *Chem. Mater.* **9,** 3159 (1997b).

Ding, J. H., and Gin, D. L., Catalytic Pd nanoparticles synthesized using a lyotropic liquid crystal polymer template, *Chem. Mater.* **12,** 22 (2000).

Doesburg, E. B. M., Hoppener, R. H., de Koning, B., Xiaoding, X., and Scholten, J. J. F., Preparation and characterization of copper/zinc oxide/alumina catalysts for methanol synthesis, *in* "Preparation of Catalysts IV" (G. Poncelet, J. Martens, B. Delmon, P. A. Jacobs, and P. Grange, Eds.), pp. 31, 767. Elsevier, Amsterdam (1987).

Ebner, K., Process and Apparatus for the Thermal Decomposition of Substances or Mixtures of Same, U.S. Patent 2,155,119, Assigned to American Lurgi Corporation (April 18, 1939).

Ebner, K., Process and Apparatus for the Thermal Decomposition of Substances or Mixtures of Same, West German Patent 753,306, Assigned to American Lurgi (1951).

Ebner, K., Thermal Decomposition of Substances, West German Patent 877,196, Assigned to American Lurgi (1953).

Emerson, S. C., Coote, C. F., Booth, H., Tufts, J. C., LaRocque, R., and Moser, W. R., The ultrasonic synthesis of nanostructured metal oxide catalysts, *in* "Studies in Surface Science and Catalysis" (G. Poncelet, J. Martens, B. Delmon, P. A. Jacobs, and P. Grange, Eds.), pp. 118, 773 (1998).

Find, J., Emerson, S. C., Krausz, I. M., and Moser, W. R., Hydrodynamic cavitation as a tool to control, macro-, micro-, and nano-properties of inorganic materials, *J. Mater. Res.* (2001).

Fuentes, S., and Figueras, F., The influence of particle size on the catalytic properties of alumina-supported rhodium catalysts. *J. Catal.* **61,** 443 (1980).

Gleiter, H., *in* "Deformation of Polycrystals: Mechanism and Microstructure" (N. Hansen, A. Horsewell, T. Leffers, and H. Lilholt, Eds.), **15,** Riso National Laboratory, Rackilde, Denmark (1981).

Gleiter, H., Nanocrystalline materials. *Prog. Mater. Sci.* **33,** 223 (1989).

Grinstaff, M. W., Cichowlas, A. A., Choe, S. B., and Suslick, K. S., Effect of cavitation conditions on amorphous metal synthesis. *Ultrasonics* **30,** 168 (1992).

Haruta, M., Size- and support-dependency in the catalysis of gold. *Catal. Today* **36,** 153 (1997).

Haruta, M., Tsubota, S., Kobayashi, T., Kageyama, H., Genet, M. J., and Delmon, B., Low temperature oxidation of CO over gold supported on titania, iron oxide, and cobalt oxide. *J. Catal.* **144,** 175 (1993).

Henry, C. R., Chapon, C., Giorgio, S., and Goyhenex, C., Size effects in heterogeneous catalysis: A surface science approach, *in* "Chemisorption and Reactivity on Supported Clusters and Thin Films," 331, 117, NATO ASI Ser., Ser. E, Dordrecht (1997).

Hyeon, T., Fang, M., and Suslick, K. S., Nanostructured molybdenum carbide: Sonochemical synthesis and catalytic properties, *J. Am. Chem. Soc.* **118,** 5492 (1996).

ICDD, *International Centre for Diffraction Data, Powder Diffraction File—2 Sets 1–47, Card 18–117* (1997a).

ICDD, *International Centre for Diffraction Data, Powder Diffraction File—2 Sets 1–47, Card 37–381* (1997b).

ICDD, *International Centre for Diffraction Data, Powder Diffraction File—2 Sets 1–47, Card 21–570* (1997c).

Jordan, M., Reznek, S. R., Neville, M. A., Soucy, B. A., and Mackay, B. E., High Surface Area Metal Oxide Foams and Method of Producing Same, U.S. Patent 4,937,062, Assigned to Cabot Corporation (June 26, 1990).

Kaldor, A., and Cox, D. M., Hydrogen chemisorption on gas-phase transition metal clusters. *J. Chem. Soc. Faraday Trans.* **86,** 2459 (1990).

Knapp, R. T., Daily, J. W., and Hammitt, F. G., "Cavitation", McGraw-Hill, New York. (1979).

Kozyuk, O. V., Method of obtaining free disperse system and device for effecting same, U.S. Patent 5,492,654, Exclusively Licensed to Five Star Technologies (Feb. 10, 1996).

Kozyuk, O. V., Method of Obtaining a Free Disperse System in Liquid and Device for Effecting the Same, U.S. Patent 5,810,052, Assigned to Five Star Technologies (Sept. 22, 1998).

Kozyuk, O. V., Method and Apparatus for Producing Ultra-thin Emulsions and Dispersions, U.S. Patent 5,931,771, Exclusively Licensed to Five Star Technologies (August 3, 1999a).

Kozyuk, O. V., Method and Apparatus For Conducting Sonochemical Reactions and Processes Using Hydrodynamic Cavitation, U.S. Patent 5,937,906, Exclusively Licensed to Five Star Technologies (August 17, 1999b).

Kozyuk, O. V., Method for Changing the Quantitative and Qualitative Composition of a Mixture of Liquid Hydrocarbons based on the Effects of Cavitation, U.S. Patent 5,969,207, Exclusively Licensed to Five Star Technologies (October 19, 1999c).

Kozyuk, O. V., Method and Apparatus of Producing Liquid Disperse Systems, U.S. Patent 5,971,601, Exclusively Licensed to Five Star Technologies (October 26, 1999d).

Kozyuk, O. V., Method and Apparatus For Conducting Sonochemical Reactions and Processes Using Hydrodynamic Cavitation, U.S. Patent 6,035,897, Exclusively Licensed to Five Star Technologies (March 14, 2000b).

Kozyuk, O. V., Method and Apparatus For Conducting Sonochemical Reactions and Processes Using Hydrodynamic Cavitation, U.S. Patent 6,012,492, Exclusively Licensed to Five Star Technologies (January 11, 2000a).

Li, S., Lee, J. S., Hyeon, T., and Suslick, K. S., Catalytic dehydronitrogenation of indole over molybdenum nitride and carbide with different structures. *Appl. Catal. A: General* **184,** 1 (1999).

Li, T., Moon, J., Morrone, A. A., Mecholsky, J. J., Talham, D. R., and Adatr, J. H., Preparation of Ag/SiO_2 nanosize composites by a reverse micelle and sol-gel technique. *Langmuir* **15,** 5328 (1999).

Martino, A., Kawola, J. S., Sault, A. G., Hampden-Smith, M., and Yamanaka, S. A., Highly dispersed pseudo-homogeneous and heterogeneous catalysts synthesized via inverse micelle solutions for the liquefaction of coal. SAND98-2630, p. 1–121, Sandia National Laboratories, Albuquerque, NM (1998).

Martino, A., Sault, A. G., Kawola, J. S., Boespflug, E., and Phillips, M. L., A sintering study of novel sol-gel nanocluster catalysts. *J. Catal.* **187,** 30 (1999).

Martino, A., Stoker, M., Hicks, M., and Bartholomew, C. H., Synthesis and characterization of Fe colloid catalysts in inverse micelle solutions. *Preprnt. Petrol. Div., Amer. Chem. Soc.* **40**(1), 78 (1995).

Martino, A., Yamanaka, S. A., Kawola, J. S., and Loy, D. A., Encapsulation of gold nanoclusters in silica materials via an inverse micelle/sol-gel synthesis. *Chem. Mater.* **9,** 423 (1997).

Masson, A., Bellamy, B., Romdhane, Y. H., Che, M., Roulet, H., and Dufour, G., Intrinsic size effect of platinum particles supported on plasma-grown amorphous alumina in the hydrogenation of ethylene. *Surf. Sci.* **173,** 479 (1986).

Matijevic, E., Control of powder morphology, *in* "Chemical Processing of Advanced Materials" (L. H. Hench and J. K. West, Eds.), p. 513. Wiley, New York (1992).

McNamara, W. B., Didenko, Y., and Suslick, K. S., Sonoluminescence temperatures during multibubble cavitation. *Nature* **401,** 772 (1999).

Mdleleni, M. M., Hyeon, T., and Suslick, K. S., Sonochemical synthesis of nanostructured molybdenum sulfide, *J. Am. Chem. Soc.* **120,** 6189 (1998).

Michalakos, P. M., Bellis, H. E., Brusky, P., Kung, H. H., Li, H. Q., Moser, W. R., Partenheimer, W., and Satek, L. C., Synthesis of vanadium phosphorus oxide catalysts by aerosol processing. *Ind. Eng. Chem. Res.* **34**(6) (1995).

Montejano-Carrizales, and Moran-Lopez, J. L., Geometrical characteristics of compact nanoclusters. *Nanostructured Mater.* **1,** 397 (1992).

Moser, W. R., Hydrodesulfurization Catalyst and Process Utilizing the Same, U.S. Patent 4,018,672, Assigned to Exxon (April 19, 1977).

Moser, W. R., Hydrodesulfurization Catalyst, U.S. Patent 4,090,982, Assigned to Exxon (May 23, 1978a).

Moser, W. R., Hydrodesulfurization of Oil Utilizing a Catalyst of Rare Earth Metal, Non-Rare Earth Metal and Alumina Support, U.S. Patent 4,092,239, Assigned to Exxon (May 30, 1978b).

Moser, W. R., Rare Earth Metal Oxide Hydrodesulfurization Catalysts, U.S. Patent 4,108,589, Assigned to Exxon (August 8, 1978c).

Moser, W. R., Reactor for High-Temperature Aerosol Decomposition and Catalysts and Other Materials Obtained, PCT Int. Appl., Assigned to Worcester Polytechnic Institute (1991).

Moser, W. R., Hydrocarbon partial oxidation catalysts prepared by the high-temperature aerosol decomposition process, *in* "Catalytic Selective Oxidation, ACS Symp. Ser." (S. T. Oyama and J. W. Hightower, Eds.), pp. 523, 244 (1993).

Moser, W. R., Preparation of Nanophase Solid State Materials, U.S. Patent 5,417,956, Assigned to Worcester Polytechnic Institute (May, 23, 1995a).

Moser, W. R., Process for the Preparation of Solid State Materials, U.S. Patent 5,466,646, Assigned to Worcester Polytechnic Institute (November 14, 1995b).

Moser, W. R., Developments in heterogeneous selective oxidation catalysis, *in* "New Developments in Selective Oxidation Catalysis," vol. 13, The Catalyst Group, Multi-client Study, Spring House, PA (1996).

Moser, W. R., and Cnossen, J. E., Aerosol propylene oxidation catalysts, *Preprnt. Pet. Chem. Div., Amer. Chem. Soc.* **37,** 1105 (1992).

Moser, W. R., and Connolly, K. E., Synthesis and characterization of copper modified zinc chromites by the high-temperature decomposition (HTAD) process for higher alcohol synthesis, *Chem. Eng. J.* **64,** 239 (1996).

Moser, W. R., Giang, T., Nguyen, S., and Kozyuk, O. V., A new route to cavitational chemistry and chemical processing by controlled flow cavitation, *in* "Process Intensification for the Chemical Industry, 3rd International Conference" (A. Green, Ed.), pp. 38, 173, BHR Group, London (1999).

Moser, W. R., Knapton, J. A., Koslowski, C. C., Rozak, J. R., and Vezis, R. H., Noble metal aerosol catalysts. *Preprnt. Pet. Chem. Div., Amer. Chem. Soc.* **38**(1), 829 (1993).

Moser, W. R., Knapton, J. A., Koslowski, C. C., Rozak, J. R., and Vezis, R. H., Noble metal catalysts prepared by the high-temperature aerosol decomposition (HTAD) process, *Catal. Today* **21,** 157 (1994).

Moser, W. R., and Lennhoff, J. D., Synthesis of mixed metal oxides for ceramics and catalysts by the HTAD process, *Proc. Mater. Res. Soc.,* Boston, MA (1984).

Moser, W. R., and Lennhoff, J. D., Synthesis of mixed metal oxides for ceramics and catalysts by the HTAD process. *Chem. Eng. Commun.* **83,** 241 (1989).

Moser, W. R., Lennhoff, J. D., Cnossen, J. E., Fraska, K., Schoonover, J. W., and Rozak, J. R., The preparation of advanced catalytic materials by aerosol processing, *in* "Advanced Catalysts and Nanostructured Materials, Novel Preparative Techniques" (W. R. Moser, Ed.), p. 535, Academic Press, New York (1996).

Moser, W. R., Lennhoff, J. D., Cnossen, J. E., Fraska, K., and Rozak, J. R., The high-temperature aerosol decomposition process: A general method for the synthesis of complex catalysts. *Preprnt. Petroleum. Div., Amer. Chem. Soc.* **40**(1), 49 (1995).

Moser, W. R., Marshik, B. J., Kingsley, J., Lemberger, M., Willette, R., Chan, A., Sundstrom IV, J. E., and Boye, A., The synthesis and characterization of solid-state materials produced by high shear hydrodynamic cavitation. *J. Mater. Res.* **10,** 2322 (1995).

Moser, W. R., Morikis, T., Zawadski, J., Sunstrom IV, J. E., and Marshik-Guerts, B. J., The synthesis of nanostructured, pure phase materials by hydrodynamic cavitation, *Third International Conference on Nanostructured Materials,* Kona, HI, 285 (1996).

Nikolajenko, V., Bosacek, V., and Danes, V., Investigation of properties of the metallic nickel surface in mixed Ni-MgO catalysts. *J. Catal.* **2,** 127 (1963).

Okuhara, T., Inumaru, K., and Misono, M., Active crystal face of vanadyl pyrophosphate for selective oxidation of *n*-butane, *in* "Catalytic Selective Oxidation" (S. T. Oyama and J. W. Hightower, Eds.), Vol. 523, p. 156, *Am. Chem. Soc. Symp. Ser.* (1993).

Salama, T. M., Hattori, H., Kita, H., Ebitani, K., and Tanaka, T., X-ray adsorption spectroscopic and electron paramagnetic resonance studies on the strong metal-support interaction of platinum supported on titania dispersed on silica, *J. Chem. Soc. Faraday Trans.* **89**(12), 2067 (1993).

Sanchez, R. M. T., Ueda, A., Tanaka, K., and Haruta, M., Selective oxidation of CO in hydrogen on gold supported on manganese oxide, *J. Catal.* **168,** 125 (1997).

Sault, A. G., Martino, A., Kawola, J. S., and Boesplug, E., Novel sol-gel based Pt nanocluster catalysts for propane dehydrogenation, *J. Catal.* **191,** 474 (2000).

Schultz, M., and Matijevic, E., Preparation and properties of nanosized PdS dispersions for electrolytic plating, *Colloids Surf.* **131,** 173 (1998).

Sellinger, A., Weiss, P. M., Nguyen, A., Lu, Y., Assink, R. A., Gong, W., and Brinker, C. J., Continuous, self-assembly of organic–inorganic nanocomposite coatings that mimic nature. *Nature* **394,** 256 (1998).

Siegel, R. W., Cluster assembled nanophase materials. *Annu. Rev. Mater. Sci.* **21,** 559 (1991).

Siegel, R. W., Nanophase materials, *in* "Encyclopedia of Applied Physics" (G. L. Trigg, Ed.), Vol. 11, p. 173, VCH Publishers, Weinheim (1994).

Siegel, R. W., and Hahn, H., Nanophase materials, *in* "Current Trends in the Physics of Materials" (M. Yussouff, Ed.), p. 403, World Scientific, Singapore (1987).

Skandan, G., Kear, B. H., Chang, W., and Hahn, H., Apparatus for making nanostructured ceramic powders and whiskers., U.S. Patent 5,514,350, Assigned to Rutgers University (May 7, 1996).

Sunstrom IV, J. E., Moser, W. R., and Marshik-Geurts, B. J., General route to nanocrystalline oxides by hydrodynamic cavitation, *Chem. Mater.* **8,** 2061 (1996).

Suslick, K. S., The mechanochemical effects of ultrasound, *Proc. Int. Conf. Mechanochem., 1st,* **1,** 43 (1993).

Suslick, K. S., Didenko, Y., Fang, M. M., Hyeon, T., Kolbeck, K. J., McNamara, W. B., Mdleleni, M., and Wong, M., Acoustic cavitation and its chemical consequences, *Phil. Trans. Royal Soc. A* **357,** 335 (1999).

Suslick, K. S., Fang, M., Hyeon, T., and Cichowlas, A. A., Nanostructured Fe-Co catalysts generated by ultrasound, *Mat. Res. Soc. Symp. Proc.* **351,** 443 (1994).

Suslick, K. S., Fang, M. M., Hyeon, T., and Mdleleni, M. M., Applications of sonochemistry to materials synthesis, *in* "Sonochemistry and Sonoluminescence" (L. A. Crum, T. J. Mason, J. Reisse, and K. S. Suslick, Eds.), p. 291, Kluwer Publishers, Dordrecht, Netherlands (1999).

Suslick, K. S., Hyeon, T., Fang, M., and Cichowlas, A., Sonochemical preparation of nanostructured catalysts, *in* "Advanced Catalysts and Nanostructured Materials: Modern Synthetic Methods" (W. R. Moser, Ed.), p. 197, Academic Press, San Diego. (1996).

Sze, C., Gulari, E., and Demczyk, B. G., Surface structure–catalytic function in nanophase gold catalysts, *in* "Nanophase and Nanocomposite Materials" (S. Komarneni, J. C. Parker, and G. J. Thomas, Eds.), Vol. 286, p. 143, *Mater. Res. Soc.* (1993).

Tompa, G. S., Skandan, G., Glumac, N., and Kear, B. H., A new flame process for producing nanopowders, *Amer. Ceram. Soc. Bull.* (October) 70 (1999).

Tschöpe, A., Schaadt, D., Birringer, R., and Ying, J. Y., Catalytic properties of nanostructured metal oxides synthesized by inert gas condensation, *Nanostr. Mater.* **9,** 423 (1997).

Tschöpe, A., Ying, J. Y., and Tuller, H. L., *Symposium B on Materials for Sensors: Functional Nanoscaled Structures,* Saarbruecken, Germany, 111 (1996).

Tsubota, S., Cunningham, D. A. H., Bando, Y., and Haruta, M., Preparation of nanometer gold strongly interacted with titania and the structure sensitivity in low-temperature oxidation of CO, *in* "Studies in Surface Science and Catalysis, Preparation of Catalysts VI" (G. Poncelet, J. Martens, B. Delmon, P. A. Jacobs, and P. Grange, Eds.), Vol. 91, p. 227, Elsevier, Amsterdam (1995).

Tsubota, S., Ueda, A., Sakurai, H., Kobayashi, T., and Haruta, M., Applications of supported gold catalysts in environmental problems, *in* "Environmental Catalysis" (J. N. Armor, Ed.), Vol. 552, p. 420, Amer. Chem. Soc., Symp. Ser. (1994).

Van Hardeveld, R., and, F. Hartog, Influence of metal particle size in nickel-on-aerosil catalysts on surface site distribution, catalytic activity, and selectivity, *Adv. Catal.* **22,** 75 (1972).

van Santen, R. A., and Niemantsverdriet, J. W., "Chemical Kinetics and Catalysis," Plenum press, New York (1995).

Volta, J. C., Desquesnes, W., Moraweck, B., and Coudurier, G., A new method to obtain supported oriented oxides: MoO_3 graphite catalysts in propylene oxidation to acrolein, *React. Kinet. Catal. Lett.* **12,** 241 (1979).

Volta, J. C., and Tatibouet, J. M., Structure sensitivity of MoO_3 in mild oxidation of propylene, *J. Catal.* **93,** 467 (1985).

Wang, C.-C., and Ying, J. Y., Sol-gel synthesis and hydrothermal processing of anatase and rutile titania nanocrystals, *Chem. Mater.* **11,** 3113 (1999).

Wilcoxon, J. P., Martino, A., Baughman, R. L., Klavetter, E., and Sylwester, A. P., Synthesis of transition metal clusters and their catalytic and optical properties, *in* "Nanophase and Nanocomposite Materials" (S. Komarneni, J. C. Parker, and G. J. Thomas, Eds.), Vol. 286, p. 131, Mater, Res. Soc. Symp. Proc. (1993).

Wilcoxon, J. P., Williamson, R. L., and Baughman, R. J., Optical properties of gold colloids formed in inverse micelles. *J. Chem. Phys.* **98,** 9933 (1993).

Williamson, G. K., and Hall, W. H., Crystallographic strain determination by X-ray diffraction, *Acta Metall.* **1,** 22 (1953).

Wold, A., Gao, Y.-M., Miller, D., Kershaw, R., and Dwight, K., Synthesis of catalytic materials by spray pyrolysis, *in* "Advanced Catalysts and Nanostructured Materials: Novel Preparative Techniques" (W. R. Moser, Ed.), p. 505, Academic Press, San Diego (1996).

Woodward, R. B., and Hoffmann, R., "The Conservation of Orbital Symmetry," Academic Press, New York (1970).

Ying, J. Y., and Tschöpe, A., Gas phase synthesis of nonstoichiometric nanocrystalline catalysts, *in* "Advanced Catalysts and Nanostructured Materials: Modern Synthetic Methods" (W. R. Moser, Ed.), p. 231, Academic Press, San Diego (1996).

Young, F. R., "Cavitation," McGraw-Hill, New York (1989).

Zarur, A. J., Hwu, H. H., and Ying, J. Y., Reverse microemulsion-mediated synthesis and structure evolution of barium hexaaluminate nanoparticles, *Langmuir* **16,** 3042 (2000).

Zarur, A. J., and Ying, J. Y., Reverse microemulsion synthesis of nanostructured complex oxides for catalytic combustion, *Nature* **403,** 65 (2000).

Zhang, Z., Wang, C.-C., Zakaria, R., and Ying, J. Y., Role of particle size in nanocrystalline TiO_2-based photocatalysts, *J. Phys. Chem. B* **102**(52), 10871 (1998).

Zhao, M., and Crooks, R. M., Dendrimer-encapsulated Pt nanoparticles, synthesis characterization, and applications to catalysis, *Adv. Mater.* **11,** 217 (1999a).

Zhao, M., and Crooks, R. M., Homogeneous hydrogenation catalysis using monodisperse, dendrimer-encapsulated Pd and Pt nanoparticles, *Angew. Chem. Int. Ed.* **38,** 364 (1999b).

SUPPORTED NANOSTRUCTURED CATALYSTS: METAL COMPLEXES AND METAL CLUSTERS

B. C. Gates

Department of Chemical Engineering and Materials Science, University of California, Davis, California 95616

I. Introduction: Supported Nanostructures as Catalysts 50
II. Supported Metal Complexes—Molecular Analogues Bonded to Surfaces 51
A. Preparation 52
B. Determination of Composition 53
C. Determination of Metal Oxidation State 53
D. Spectroscopic and Theoretical Characterization of Structure 54
E. Examples 54
F. Generalizations about Structure and Bonding 62
G. Generalizations about Reactivity and Catalysis 62
III. Metal Pair Sites and Triplet Sites on Supports 63
IV. Supported Metal Nanoclusters 64
A. Preparation 65
B. Structural Characterization 67
C. Examples 68
D. Catalytic Properties 70
E. Generalizations about Structure, Bonding, Reactivity, and Catalysis 73
V. Supported Metal Nanoparticles 73
References 74

Many catalysts are nanomaterials dispersed on the surfaces of solid supports, typically nanoporous materials. The simplest and most thoroughly investigated supported catalysts are metal complexes such as $Rh(CO)_2$ *and metal clusters such as* Ir_4*. Metals in the catalysts are covalently bonded to oxygen atoms at the surfaces of metal oxide and zeolite supports; the supports are monodentate or polydentate ligands. When the metals are from Group 7 or Group 8, the metal–oxygen bonding distances are 2.15* $\pm$ *0.10 Å, essentially the same as*

ADVANCES IN CHEMICAL ENGINEERING, VOL. 27
0065-2377/01 $35.00

in molecular analogues, as shown by extended X-ray absorption fine-structure spectroscopy and density functional theory. The syntheses and patterns of reactivity of supported metal complexes and nano-clusters are consistent with the rules of conventional organometallic chemistry, but site isolation of the supported species stabilizes coordinative unsaturation of the metals and allows for new reactivities and catalytic activities, including those for relatively unreactive compounds such as alkanes. Density functional theory indicates that supported Ir_4 clusters on zeolite NaX are nearly neutral. The theory also shows that Os_5C clusters are bonded much more strongly at surface defect sites than at defect-free sites on MgO. Supported clusters of only a few metal atoms have catalytic properties different from those of larger supported nanoclusters and nanoparticles of metal; they offer intriguing prospects for new catalytic properties.

I. Introduction: Supported Nanostructures as Catalysts

Catalysts were some of the first nanostructured materials applied in industry, and many of the most important catalysts used today are nanomaterials. These are usually dispersed on the surfaces of supports (carriers), which are often nearly inert platforms for the catalytically active structures. These structures include metal complexes as well as clusters, particles, or layers of metal, metal oxide, or metal sulfide. The solid supports usually incorporate nanopores and a large number of catalytic nanoparticles per unit volume on a high-area internal surface (typically hundreds of square meters per cubic centimeter). A benefit of the high dispersion of a catalyst is that it is used effectively, because a large part of it is at a surface and accessible to reactants. There are other potential benefits of high dispersion as well—nanostructured catalysts have properties different from those of the bulk material, possibly including unique catalytic activities and selectivities.

Besides being small, most catalytic nanomaterials are nonuniform and thus are difficult to characterize well. Consequently, much remains to be learned about them, and chances are excellent that many nanomaterials with novel and useful catalytic properties remain to be discovered.

The goal of this chapter is to illustrate supported catalytic nanomaterials, with an emphasis on those that are simple and well-defined structurally (and thus relatively well understood), and to summarize generally important conclusions about their structure, bonding, reactivity, and catalytic properties. This summary is not exhaustive, and references to related reviews are cited.

Supported catalysts may be classified according to the dimensions of the catalytic components, as follows:

- Mononuclear (single-metal-atom) complexes that are closely analogous to molecular structures
- Multiatom metal, metal oxide, metal sulfide, metal carbide, metal nitride, or metal phosphide nanoclusters that are so small that they have properties different from those of the corresponding bulk materials
- Nanolayers of these materials, the most widely investigated being metal oxides
- Nano- (or larger) particles of these materials that have bulklike properties

This classification includes a distinction between nanoclusters and nanoparticles, although there is still no generally accepted usage distinguishing these terms.

The emphasis here is principally on metals rather than the other materials, because metals are simpler, more widely investigated, and better understood than the others. Supported metal nanoparticles are considered here in only a few summary statements, and nanolayers are beyond the scope of the chapter.

II. Supported Metal Complexes—Molecular Analogues Bonded to Surfaces

The simplest supported catalysts are mononuclear metal complexes, exemplified by industrial supported metallocene catalysts, used (with promoters) for alkene polymerization; these are the so-called single-site catalysts that are finding wide industrial applications (Kristen, 1999; Kaminsky, 1999; Roscoe *et al.,* 1998). The most common supports are metal oxides and zeolites. The metals in these complexes range from oxophilic (e.g., Zr and Ta) to noble (e.g., Rh). Supported metal complexes are stabilized by ligands—in addition to those provided by the support—such as hydride (H), hydrocarbons, and carbonyl (CO). In a typical supported metal complex, the metal is present in a positive oxidation state. Although some such complexes are relatively stable, most are, befitting their roles as catalysts, highly reactive and air- and moisture-sensitive.

In the following paragraphs, methods of preparation and characterization of structurally simple supported metal complexes are summarized, and examples are presented that illustrate characterization data and support general conclusions about structure, bonding, reactivity, and catalysis.

A. Preparation

1. Organometallic Chemistry

Many syntheses of supported metal complexes involve simply the reaction of an organometallic precursor in an organic solvent with a support surface; alternatively, a gas-phase precursor may be used in the absence of a solvent. The surface reactions are typically analogous to molecular reactions known from solution organometallic chemistry. The surface chemistry has been reviewed (Lamb, Gates, and Knözinger, 1988; Basset, Lefebvre, and Santini, 1998), and only a few examples are given here.

Deprotonation of a metal hydride precursor may give a surface-bound metal complex and an OH group formed from an oxygen atom on the support surface. For example, $[H_2Os(CO)_4]$ may react with surface oxygen atoms to give surface OH groups and $[HOs(CO)_4]^-$ bonded (ion-paired) to the surface (Lamb, Gates, and Knözinger, 1988). However, the chemistry is often more complex than simple deprotonation and formation of a group (such as an anion) that becomes bonded to the support. This is acid–base chemistry, occurring more readily on a basic surface, such as that of MgO, than on a less basic surface, such as that of SiO_2.

Highly reactive metal complexes with ligands such as alkyls or allyls often react readily with oxygen atoms or OH groups of oxide or zeolite surfaces. For example, $Ta(-CH_2CMe_3)_3(=CHCMe_3)$ (Me is methyl) reacts with surface oxygen atoms of SiO_2 to form mixtures of $\{SiO-\}Ta(-CH_2CMe_3)_2(=CHCMe_3)$ (≈65%) and $\{SiO-\}_2Ta(-CH_2CMe_3)(=CHCMe_3)$ (≈35%) (Vidal *et al.,* 1997). The group denoted in braces, {SiO–}, represents part of the support—the termination of the bulk solid. {SiO–} is a monodentate ligand, and $\{SiO-\}_2$ is a bidentate ligand. In this synthesis, $(-CH_2CMe_3)$ groups are replaced by oxygen atoms of the surface.

In other preparations, oxidative addition of an O–H ligand of an oxide support surface to a metal center of an organometallic complex leads to anchoring of the metal complex to the support (Lamb, Gates, and Knözinger, 1988).

Other synthetic surface reactions are reviewed elsewhere (Lamb, Gates, and Knözinger, 1988; Basset, Lefebvre, and Santini, 1998).

2. Coordination Chemistry

In contrast to the foregoing examples, preparation chemistry on a support surface is often analogous to coordination chemistry in aqueous solutions rather than to organometallic chemistry in organic solutions. Aqueous chemistry is much more typical of the preparation of industrial catalysts than organometallic chemistry, but it is also often much more complex, involving, for example, dissolution of the support and reprecipitation of structures, including mixed-metal species (Che, 2000; Che, Clause, and Marcilly, 1999). Industrial preparation of a supported catalyst often involves a sequence of steps such as the following: contacting of the support with an aqueous solution containing a catalyst precursor such as a metal salt (impregnation), removal of the solvent water, drying, calcination, and possibly reduction and/or sulfidation.

Che (2000) identified a sequence of transformations that are commonly involved during these steps in the preparation of an oxide-supported catalyst in an aqueous environment. First, a hydrated ionic species such as $[PtCl_5(H_2O)]^-$, formed from a precursor metal salt, forms a weakly held outer-sphere complex with the surface of the support, such as γ-Al_2O_3; the interaction may be largely electrostatic, and the chemistry depends on the support and the solution properties, such as the pH. This surface species reacts to form an inner-sphere complex as surface groups (e.g., oxygen atoms) of the support enter the first coordination sphere of the metal (i.e., bond to it); thus, the support acts as a monodentate or polydentate donor ligand. Processes of dissolution–reprecipitation may occur simultaneously with this chemistry, complicating matters; these processes also depend on the solvent chemistry (e.g., the pH). Subsequent treatments, such as calcination, may convert the resultant molecular analogues on the support surface into larger, more complex, and typically nonuniform structures. There are only a few examples of characterization of the intermediates that may be regarded as molecular analogues (Che, 2000).

B. Determination of Composition

Temperature-programmed desorption (or decomposition) with quantitative analysis of gas-phase products (usually by mass spectrometry) has been used to help identify the ligands bonded to a metal in a supported complex. Complications such as reaction of desorbed ligands (e.g., CO) with support groups (e.g., OH) may complicate interpretation of the data (Brenner, 1986).

C. Determination of Metal Oxidation State

Analysis of gas-phase products formed as a precursor reacts with a support surface may provide evidence of the oxidation state of the metal in the resultant supported complex, as illustrated by the following hypothetical stoichiometry of a synthesis on a hydroxylated oxide support (Brenner, 1986):

$$M_1(CO)_j + n\{M_2 - OH\} \rightarrow \{M_2 - O^-\}_n M_1^{n+} + (n/2)H_2 + jCO. \quad (1)$$

(M_1 and M_2 refer to the metals in the precursor metal carbonyl and in the oxide support, respectively.) Temperature-programmed-oxidation and -reduction experiments provide complementary information.

D. Spectroscopic and Theoretical Characterization of Structure

Spectroscopic methods have been used often to characterize supported metal complexes. Supported metal carbonyls are among the most thoroughly investigated and provide many of the best-understood examples, in part because the CO ligands offer informative fingerprints in the IR spectra, which are easily measured. Metal carbonyl complexes on surfaces are also well suited to characterization by extended X-ray absorption fine-structure (EXAFS) spectroscopy, because the multiple scattering in the linear M–C–O groups allows one to distinguish the C and O backscatterers from other backscatterer ligands bonded to the metal, including those of the support. Supported metal complexes with organic ligands are well suited to characterization by ^{13}C NMR spectroscopy and those with hydride ligands to characterization by ^{1}H NMR spectroscopy. Other useful characterization methods include Raman and ultraviolet/visible spectroscopies.

In recent years researchers have begun to use density functional theory to model supported metal complexes, representing the support as a cluster (fragment of the bulk support). Examples of supported complexes characterized by both theory and spectroscopy are presented below.

When samples incorporate uniform mononuclear metal complexes, then EXAFS data may provide high-quality information about the interactions between the metal and oxygen atoms of the support. Evidence of the metal–support interface has also been determined by theory and, indirectly, by IR spectra indicating the symmetry and thus the number of atoms of the support that act as ligands bonded to the metal.

E. Examples

1. Rhodium Dicarbonyls on Dealuminated Y Zeolite

Rhodium carbonyls on oxide and zeolite supports have been investigated extensively and are some of the best-understood supported metal complexes. These can be formed, for example, from rhodium salts such as $[Rh(NH_3)_5Cl][OH]_2$ (Miessner *et al.*, 1989), from organometallic precursors such as $[Rh(CO)_2(acac)]$ (Goellner, Gates *et al.*, 2000), and even from supported nanoparticles of rhodium metal, which undergo oxidative fragmentation upon treatment with CO when present on hydroxylated surfaces such as that of γ-Al_2O_3 (van't Blik *et al.*, 1985).

A rhodium dicarbonyl formed from $[Rh(NH_3)_5Cl][OH]_2$ (Miessner *et al.*, 1989) or from $[Rh(CO)_2(acac)]$ (Goellner, Gates *et al.*, 2000) on dealuminated Y zeolite has been investigated with IR (Miessner *et al.*, 1989; Goellner, Gates *et al.*, 2000) and EXAFS (Goellner, Gates *et al.*, 2000) spectroscopies and density functional theory (Goellner, Gates *et al.*, 2000) (Fig. 1). The supported complex is represented as $Rh^+(CO)_2$, and the results indicate that it is bonded near an Al atom at a four-ring in the zeolite (Fig. 1). This is an example of a well-defined metal complex on a relatively simple and well-defined (crystalline) support, a faujasite containing only a low density

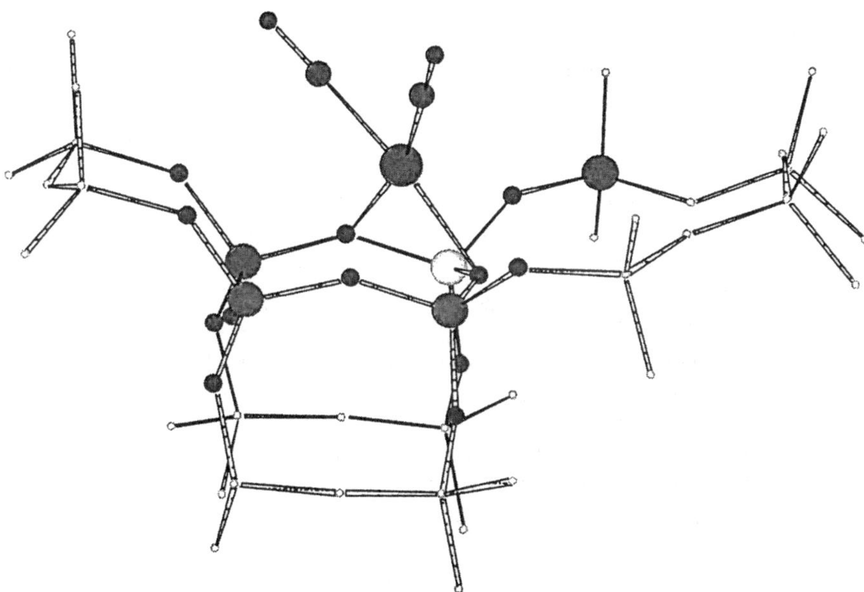

FIG. 1. Model of rhodium dicarbonyl complex on dealuminated Y zeolite, as determined by IR and EXAFS spectroscopies and density functional theory. The Rh atom, near the upper center of the figure, has two CO ligands bonded to it, pointing upward, and two oxygen atoms of the zeolite lattice below. An Al atom is located between these two oxygen atoms. The dangling atoms of the cluster model of the zeolite are capped by hydrogen atoms for the calculation (Goellner, Gates, *et al.*, 2000).

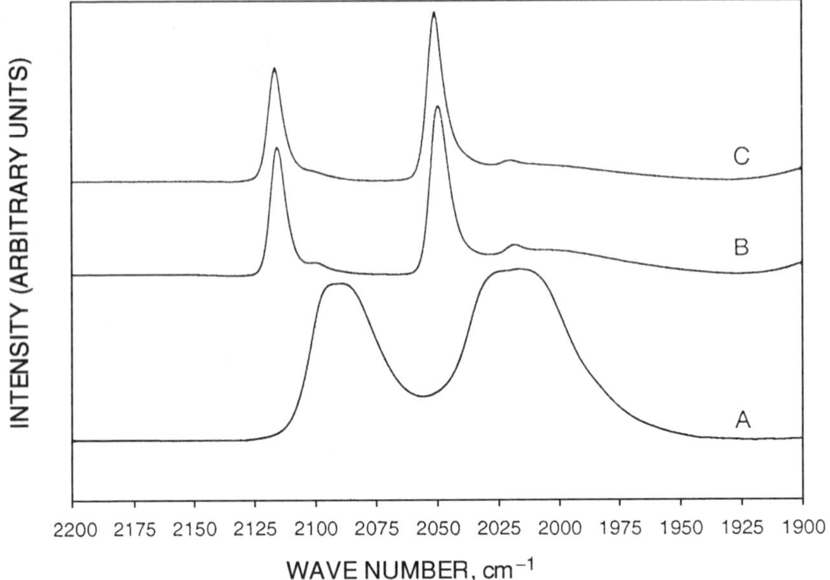

FIG. 2. Infrared spectra in the carbonyl stretching region of Y zeolite-supported $Rh(CO)_2$. (A) Complex in NaY zeolite calcined at 200°C. (B) Complex in dealuminated Y zeolite calcined at 120°C. (C) Complex in dealuminated Y zeolite calcined at 300°C (Goellner, Gates, *et al.*, 2000).

of Al atoms and hence only a low density of bonding sites for the cationic complex. The IR spectra in the carbonyl stretching region (Fig. 2, spectra B and C) are sharp and suggestive of a single species bonded to the support, in contrast to those of $Rh^+(CO)_2$ on a zeolite (NaY) with a higher Al : Si ratio (Fig. 2, spectrum A). The EXAFS spectra and predictions of density functional theory (Table I) agree well with each other and with the IR spectra, being consistent with the structure of Fig. 1. There are two Rh–O bonds in the complex of Fig. 1, and the Rh–O distances are about 2.2 Å (Table I). These are covalent bonding distances, similar to those observed by X-ray diffraction (XRD) crystallography for pure compounds with Rh–O bonds. It is clear that the zeolite support is a bidentate ligand and that the bonds between Rh and O are ordinary covalent bonds.

Miessner (1994) showed that partial decarbonylation of this supported complex by treatment in H_2 at temperatures of 200 to 250°C leads to complexes that are so highly reactive that they combine with N_2 to give well-defined supported complexes with dinitrogen ligands. This remarkable reactivity suggests possibilities for new catalytic properties of these and related supported metal complexes.

TABLE I

COMPARISON OF STRUCTURAL PARAMETERS DETERMINED BY DENSITY FUNCTIONAL THEORY AND BY EXAFS SPECTROSCOPY CHARACTERIZING SAMPLES MADE FROM CHEMISORPTION OF $[Rh(CO)_2(acac)]$ ON DEALUMINATED Y ZEOLITE (GOELLNER, GATES *et al.*, 2000)[a]

Calcination temperature (°C) of zeolite support	Calculated			Experimental (EXAFS)				
	Contribution	N	R (Å)	Backscatterer	N	R (Å)	$10^3 \cdot \Delta\sigma^2$(Å^2)	ΔE_0(eV)
300				CO				
	Rh–C	2	1.88	C	2.2	1.86	–0.96	–16.45
	Rh–O*	2	3.03	O*	2.3	2.96	0.49	–7.47
				support				
	Rh–O	2	2.19–2.20	O_{short}	1.9	2.15	0.61	–9.42
	Rh–Al	1	2.80	Al	1.1	2.73	0.88	–17.29
120				CO				
	Rh–C	2	1.88	C	2.3	1.86	–1.33	–17.17
	Rh–O*	2	3.03	O*	2.2	2.97	0.69	–7.80
				support				
	Rh–O	2	2.19–2.20	O_{short}	1.8	2.16	–0.32	–9.30
	Rh–Al	1	2.80	Al	1.3	2.74	0.38	–19.71

[a]Notation: N, coordination number; R, distance between absorber and backscatterer atom; $\Delta\sigma^2$, Debye–Waller factor; ΔE_0, inner potential correction. Commonly accepted error bounds on structural parameters obtained by EXAFS spectroscopy are N, ±10–15%; R, ±0.02 Å; $\Delta\sigma^2$, ±20%; and ΔE_0, ±20%.

2. Rhodium Dicarbonyls on γ-Al_2O_3

Rhodium dicarbonyls on γ-Al_2O_3, similar to those on dealuminated zeolite Y, have been investigated extensively, as summarized by Wovchko and Yates (1999). These complexes, which presumably can be represented as $[\{AlO\}_2Rh(CO)_2]$, can be activated photochemically to remove CO ligands. The resultant surface species react with N_2 in chemistry similar to that mentioned previously for the zeolite-supported analogue; they also react with methane, leading to formation of methyl ligands, which undergo insertion reactions with CO ligands (Wovchko and Yates, 1999). The results suggest possibilities for new catalysts for alkane conversion.

3. Rhenium Tricarbonyls on MgO

The analogy between molecular and supported metal complexes is strongly reinforced by results representing rhenium tricarbonyls on MgO. These are some of the most thoroughly investigated samples in this category, having been characterized with many experimental methods as well as density functional theory. The complexes were made on MgO powder from various organometallic precursors, $[HRe(CO)_5]$, $[DRe(CO)_5]$, $[Re_2(CO)_{10}]$, and $[H_3Re_3(CO)_{12}]$ (Kirlin *et al.,* 1990a, 1990b; Triantafillou, Purnell *et al.,* 1994); they have also been made on ultrathin MgO films exposing the (111) face (Purnell *et al.,* 1994a,b). The experimental methods used to investigate the samples were temperature-programmed decomposition and reduction (which confirm the composition) and IR, EXAFS, Raman, ultraviolet-visible, and inelastic electron tunneling spectroscopies (Kirlin, De Thomas *et al.,* 1986; Triantafillou, Purnell *et al.,* 1994). The EXAFS data indicate approximately three CO ligands per Re atom and approximately three support oxygen atoms neighboring each Re atom, consistent with the vibrational spectra and the other data (Papile and Gates, 1992).

Rhenium carbonyls have been prepared on MgO powder with various degrees of hydroxylation (and methoxylation), and IR and EXAFS spectra indicate symmetries of the complexes, e.g., C_{3v} on dehydroxylated MgO; this complex is represented as $Re(CO)_3\{OMg\}_3$ (Papile and Gates, 1992). The IR data show substantial shifts in the ν_{co} bands with changes in the degree of hydroxylation or methoxylation of the MgO surface (Table II), even providing a basis for determining the number of O versus OH ligands from the support that are bonded to a Re atom.

Two limiting-case surface structures, $Re(CO)_3\{OMg\}_3$ and $Re(CO)_3\{HOMg\}_3$, represented here as structures A and B, have been modeled with density functional theory, with the MgO being represented as a fragment of its surface, shown in the structures below (Hu *et al.,* 1999). Consistent with the experimental results, the most stable species (structure A) was found to exist

TABLE II

SUMMARY OF STRUCTURAL ASSIGNMENTS AND IR SPECTRA OF RHENIUM TRICARBONYLS ON MgO (PAPILE AND GATES, 1992)

Structural model	Precursor	MgO pretreatment temperature, °C	Symmetry indicated by IR spectrum	v, frequency of high-frequency carbonyl band, cm^{-1}	Δv,[a] cm^{-1}	$\Delta v/(\Delta v)_{max}$	$N_{O^{2-}}/(N_{O^{2-}} + N_{OH^-})$[b]
$Re(CO)_3\{OMg\}_3$	$[Re_2(CO)_{10}]$	700	C_{3v}	2036	22	1.00	1.00
$Re(CO)_3\{OMg\}_2\{HOMg\}$	$[H_3Re_3(CO)_{12}]$	700	C_s	2028	14	0.64	0.67
$Re(CO)_3\{OMg\}\{HOMg\}_2$	$[Re_2(CO)_{10}]$	400	C_s	2022	8	0.36	0.33
$Re(CO)_3\{HOMg\}_3$	$[Re_2(CO)_{10}]$	390	C_{3v}	2014	0	0.00	0.00

[a] Change in frequency of CO vibration relative to that of $Re(CO)_3\{HOMg\}_3$.

[b] Fraction of groups on MgO surface estimated to be oxygen atoms rather than OH groups.

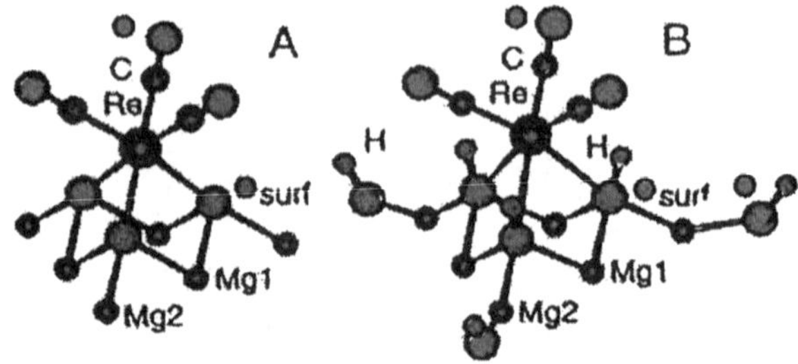

at Mg cation vacancies (corner sites) at the MgO surface. The calculated results agree very well with the symmetry indicated by the IR and EXAFS data; the predictions of the carbonyl stretching frequencies are also in good agreement with experiment (Table III). The theoretical results show that the Re–O bond energy in the dehydroxylated sample (3.5 eV) is greater than the Re–CO bond energy (2.4–2.5 eV), which reinforces the view of the support as a strongly bound tridentate ligand (Table III). The theoretical results (Hu *et al.*, 1999) indicate that the Re–$O_{support}$ distance in $Re(CO)_3\{OMg\}_3$ is 2.15 Å, which is the same as the EXAFS value observed for rhenium carbonyls on MgO powders with few surface OH groups (Triantafillou, Purnell *et al.*, 1994), approximated as $Re(CO)_3\{OMg\}_2\{HOMg\}$. The results imply that the most stable of the surface species, $Re(CO)_3\{OMg\}_3$, is formed preferentially, for example, over $Re(CO)_3\{HOMg\}_3$ or $Re(CO)_3\{OMg\}_2\{HOMg\}$, on slightly hydroxylated MgO surfaces.

TABLE III

CALCULATED BOND DISTANCES r, BINDING ENERGY TO THE SUPPORT SURFACE, E_b, AND FREQUENCIES OF $Re(CO)_3$/MgO BONDED TO MgO AT VARIOUS ASSUMED DEFECT SITES: COMPARISON OF PREDICTIONS OF DENSITY FUNCTIONAL THEORY (HU *et al.*, 1999) AND EXPERIMENTAL RESULTS (TRIANTAFILLOU *et al.*, 1994)[a]

Sample	r(Re–O)	r(Re–C)	r(C–O)	E_b	ν(Re–MgO)	ν(Re–CO)	$\Delta\nu$(C–O)[b]
$Re(CO)_3\{OMg\}_3$							
Theory							
$Re(0)(CO)_3/V_s^{-c}$	2.26	1.90	1.18	2.79	528	510	−153
$Re(I)(CO)_3/V_s^{-c}$	2.15	1.95	1.16	3.51	552	479	−93
$Re(I)(CO)_3/V_s^{c}$	2.05	2.02	1.15	2.74	551	439	−29
Experiment[d]	2.15	1.88					−107

[a] Distances r in Å, binding energy E_b of $Re(CO)_3$ species to MgO in eV per Re–O bond (three bonds), vibrational frequencies ν in cm^{-1}.

[b] Frequency shift with respect to free CO: experiment, 2143 cm^{-1}; calculation, 2120 cm^{-1}.

[c] Notation: Re(0) is zerovalent rhenium, Re(I) is Re^+. V_s refers to a cation defect site; removal of Mg^+ from the lattice gives V_s^-, for example.

[d] Further details given elsewhere (Triantafillou *et al.*, 1994).

In summary, the theoretical results representing the rhenium carbonyls on MgO are in very good agreement with the symmetry indicated by the vibrational spectra and with the coordination numbers and bond distances indicated by EXAFS spectroscopy. Thus, the rhenium carbonyls are regarded as prototype supported metal complexes. They reaffirm the strong analogy between surface-bound metal complexes and molecular metal complexes. The MgO surface is clearly identified as a polydentate ligand.

4. *Osmium Di- and Tricarbonyls on γ-Al_2O_3*

IR and EXAFS data representing osmium carbonyls on γ-Al_2O_3 led to the identification of osmium di- and tricarbonyls and of the conversion of the latter into the former. The loss of one CO ligand from the tricarbonyl complex (indicated by infrared spectroscopy and confirmed approximately by EXAFS spectroscopy) was accompanied by an increase in the Os–O coordination number from 2.9 to 3.9 (or nearly from 3 to 4) (Deutsch, Chang, and Gates, 1993), as follows:

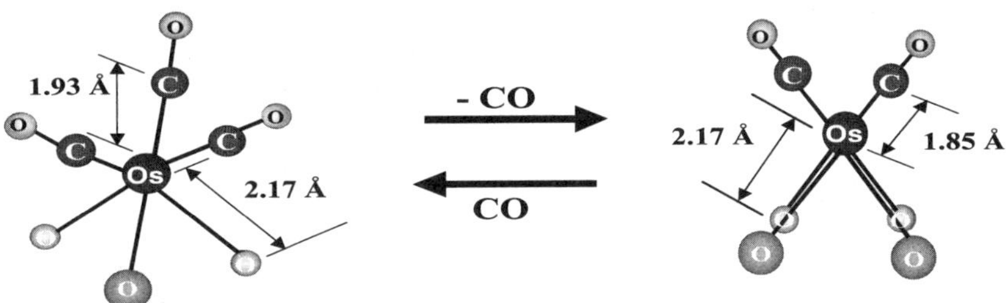

(The bond distances were determined by EXAFS spectroscopy (Deutsch, Chang, and Gates, 1993).) Thus, one CO ligand was replaced by one O ligand of the support surface, and both supported complexes were coordinatively saturated (18-electron) complexes (with O of the surface regarded formally as a two-electron donor). The Os–O, Os–C, and Os–O* distances (O* is carbonyl oxygen) are nearly equal to those of other supported Group 7 and Group 8 metal carbonyls and their molecular analogues.

These results strengthen the analogy between molecular and supported metal carbonyl complexes and extend it beyond structure and bonding to reactivity.

Osmium carbonyls on MgO and on γ-Al_2O_3, among other oxides, are catalysts (or catalyst precursors) for alkene isomerization and hydrogenation (Li *et al.,* 1984). The activity depends on the metal oxide used as a support. The ligands present on the metal during catalysis have not yet been elucidated.

5. *Tantalum Hydride Complexes on SiO_2*

When the SiO_2-supported tantalum complexes mentioned previously were treated with H_2, a surface tantalum hydride formed, $HTa\{OSi\}_2$. This catalyzes propane metathesis, giving equimolar mixtures of ethane and butane (Vidal *et al.*, 1997). This is a new reaction, clear evidence of new catalytic properties associated with a highly reactive surface-bound metal complex. In this example—and the examples involving rhodium complexes reacting with N_2 and with CH_4, for example—it is important that the rather rigid polydentate ligand provided by the support surface stabilizes the metals with degrees of coordinative unsaturation that readily allow bonding of even relatively unreactive compounds—and sometimes subsequent catalytic conversion. This is an advantage of site isolation of the metal complex; analogous metal complexes in solution do not easily retain coordinative unsaturation because they react with each other.

F. Generalizations about Structure and Bonding

The results summarized here support the following generalizations about structure and bonding of metal complexes bonded to metal oxide and zeolite supports:

- The metal–oxygen distances in metal oxide–supported complexes of Group 7 and Group 8 metals are 2.15 ± 0.10 Å.
- These distances are essentially the same as the metal–oxygen distances determined by XRD crystallography for complexes of the same metals containing electron donor oxygen ligands.
- Metal oxide and zeolite supports are ligands (sometimes monodentate and sometimes polydentate), and the metals are covalently bonded to the oxygen atoms of the support surface.
- The results of density functional theory (for supported rhodium carbonyls and rhenium carbonyls) match the available EXAFS and infrared data well.
- Most reports of supported metal complexes lack evidence of the support ligands, but we may infer that they are usually oxygen atoms or OH groups of the surface when the support is a metal oxide or a zeolite.

G. Generalizations about Reactivity and Catalysis

- The patterns of reactivity typically mimic those of analogous metal complexes in solution; the rules of organometallic chemistry apply.

- However, the fact that the supported metal complexes are site-isolated (and unable to react with each other) opens up possibilities for stabilization of coordinative unsaturation of the metals and new reactivities.
- Some supported metal complexes have new catalytic properties, and new supported metal complexes with unexpected catalytic properties seem likely to be discovered.

III. Metal Pair Sites and Triplet Sites on Supports

The metal complexes described previously, except for those supported on the dealuminated zeolite, were bonded to intrinsically heterogeneous surfaces, and the bonding sites have not been determined precisely. To a first approximation, the metal complexes on some supports (e.g., SiO_2 or γ-Al_2O_3) might be assumed to be almost randomly distributed over the surfaces, but on other supports (e.g., MgO), this seems less likely. Much remains to be learned about the specific sites on the supports where the complexes form, specifically, whether these are defect sites.

An obvious question is whether patterned arrays of metal complexes can be formed on supports, and an approach to the preparation of such materials has been made by use of precursors containing more than one metal atom. Thus, attempts have been made to prepare supported metals with pair (and triplet) sites from dimeric (and trimeric) complexes of oxophilic metals, including Mo, W, and Re, which bond strongly to oxide surfaces.

EXAFS data characterizing SiO_2-supported molybdenum species made from $[Mo_2(allyl)_4]$ have led to precise structural models, including those of surface dimers (with Mo–Mo bonds) and pair sites (without Mo–Mo bonds) (Iwasawa, 1987), but the structures seem to be sensitive to undefined chemistry of the SiO_2 support surfaces and could be difficult to reproduce. Catalytic data for ethylene hydrogenation and butadiene hydrogenation with these samples and with samples expected to have isolated Mo sites point to a catalytic role of the neighboring sites (Iwasawa, 1987).

SiO_2-supported tungsten species prepared from the dimer $[Cp_2W_2(CO)_4]$ are characterized by EXAFS data indicating a W–W contribution at a distance of 2.96 Å (too long for a W–W bond); the W–W contribution was found to be small, and stronger evidence of the presence of tungsten pair sites on the support surface is provided by reaction data for the McMurry-type reductive coupling of acetaldehyde or acetone, which was observed in the presence of the sample made from the dimeric precursor but not in the presence of a sample made similarly from a monomeric precursor (Sullivan *et al.*, 1995).

A family of catalysts was made from the aforementioned rhenium carbonyl precursors on MgO, including $[HRe(CO)_5]$, $[Re_2(CO)_{10}]$, and $[H_3Re_3(CO)_{12}]$ (Kirlin, Knözinger, and Gates, 1990; Kirlin, van Zon *et al.*, 1990; Triantafillou *et al.*, 1994). EXAFS data characterizing the samples made from $[H_3Re_3(CO)_{12}]$ indicate Re–Re contributions at an average (nonbonding) distance of 3.94 Å (Kirlin, van Zon, *et al.*, 1990). The catalyst made from $[HRe(CO)_5]$ was found to lack such contributions and to be catalytically active for propene hydrogenation but not cyclopropane hydrogenolysis, whereas the catalyst prepared similarly from $[H_3Re_3(CO)_{12}]$ is catalytically active for both reactions, as are supported nanoparticles of metallic rhenium (Kirlin and Gates, 1987; Kirlin, van Zon, *et al.*, 1990). This comparison suggests that the catalyst made from the mononuclear precursor had isolated rhenium centers, whereas the catalyst made from the trinuclear precursor had neighboring rhenium centers (perhaps groups of three) that constitute the catalytic sites for cyclopropane hydrogenolysis.

These results seem to offer a foretaste of possibilities for tailoring multicenter catalytic sites on support surfaces, but this field is still in its infancy.

IV. Supported Metal Nanoclusters

The importance of supported metal nanoclusters and nanoparticles in catalysis and the rough analogy between supported nanoclusters and organometallic cluster compounds (those with metal–metal bonds) in catalysis have motivated researchers to find connections between these two classes of materials. Thus, an obvious synthetic goal has been size-selected metal nanoclusters on supports.

Supported metal nanoclusters differ from the supported metal complexes discussed in the earlier part of this chapter by having more than one metal atom and metal–metal bonds; they may or may not have stabilizing ligands in addition to the support, and when they lack such ligands they have coordinative unsaturation and can bond to reactive ligands and be catalytically active. Part of the motivation for attempting to prepare uniform (size-selected) nanoclusters is related to the sometimes dramatic changes in properties of materials that result as the size decreases from bulk to molecular (nanocluster) dimensions. Thus, one thinks of tuning the catalytic properties of these clusters by size selection.

This section of this chapter includes a brief review of methods of preparation and properties of supported metal nanoclusters; only catalysts that have been relatively well characterized and found to be nearly uniform are considered. The nanoclusters described here lack the structural definition of organometallic cluster compounds such as metal carbonyl clusters (e.g., $[Rh_6(CO)_{16}]$), and, by and large, they are not as well defined structurally as the simpler supported mononuclear metal complexes described here. The supported clusters described here differ from bare metal nanoclusters in the gas phase, because the supports provide ligands. Nonetheless, supported nanoclusters may be coordinatively unsaturated by virtue of their stable isolation, so that reactants can bond to them.

A. Preparation

In attempts to prepare uniform supported clusters on well-defined planar supports, physicists have relied on impingement of beams of size-selected gas-phase clusters (Meiwes-Broer, 2000; Heiz *et al.,* 1999; Heiz and Schneider, 2000). This method has become quite popular, with work now underway in a number of laboratories, and it offers the advantage that, in prospect, it can be used with any metal of any cluster size. Limitations are the restriction to (nearly) planar supports (because of the preparation method) and the uncertainty of the cluster size distributions after deposition. Heiz and colleagues (1999) asserted that "soft landings" of clusters on a support lead to uniform site-isolated clusters, but recent scanning tunneling microscopy images of Si clusters on Ag(111) (Messerli *et al.,* 2000) indicate that some migration and aggregation (sintering) of the clusters is to be expected, at least under some conditions.

An alternative to this physical method of preparing structurally uniform metal clusters on supports involves chemistry by which molecular metal carbonyl clusters (e.g., $[Rh_6(CO)_{16}]$) serve as precursors on the support. These precursors are decarbonylated with maintenance of the metal frame to give supported nanoclusters (e.g., Rh_6). Advantages of this chemical preparation method are its applicability to many porous supports, such as zeolites (and not just planar surfaces) and the opportunities to use spectroscopic methods to follow the chemistry of synthesis of the precursor on the support and its subsequent decarbonylation. Zeolites, because their molecular-scale cages are part of a regular (crystalline) structure, offer the prospect of regular three-dimensional arrays of nanoclusters.

A limitation of supported metal nanoclusters prepared from molecular metal carbonyl clusters is that, so far, clusters of only several metals (Ru, Rh, Ir, and Os) have been made in high yields (80 to 90%, with the likely impurity species being mononuclear metal complexes). However, this disadvantage is offset by the advantage of the characterizations, which show that some clusters are stable even during catalysis, at least under mild conditions.

Supported nanoclusters made from metal carbonyl clusters are emphasized here, because there are numerous characterization data on which to base the discussion. The synthetic methods are illustrated by the following examples.

Molecularly or ionically dispersed metal carbonyl clusters on metal oxides have been prepared in high yields by reaction of metal carbonyl clusters with support surfaces or by syntheses on support surfaces from mononuclear precursors (Gates and Lamb, 1989; Iwasawa, 1993; Ichikawa, 1992; Gates, 1994). Synthesis of supported metal carbonyl clusters has been reviewed recently (Gates, 1995, 1998), and only a few examples are included here.

A simple reaction involves a metal carbonyl cluster and surface OH groups. For example, $[Ir_4(CO)_{12}]$ reacts with OH groups of MgO, giving adsorbed $[HIr_4(CO)_{11}]^-$ + HCO_3^-. The adsorbed $[HIr_4(CO)_{11}]^-$ is converted into $[Ir_6(CO)_{15}]^{2-}$ by treatment in CO (Zhao and Gates, 1997).

Another class of synthesis reaction is deprotonation of a hydrido metal carbonyl cluster on a basic surface. For example, $[H_4Os_4(CO)_{12}]$ reacts with MgO or with γ-Al_2O_3 to give $[H_3Os_4(CO)_{12}]^-$, which is part of a surface ion pair on the support (Budge, Scott, and Gates, 1983).

Some metal carbonyl clusters can be prepared efficiently on surfaces from mononuclear metal carbonyl precursors, e.g., $[Rh(CO)_2(acac)]$, $[Ir(CO)_2(acac)]$, and $[Pt(allyl)_2]$. The chemistry occurring on γ-Al_2O_3 illustrates the subtlety of the syntheses and the analogy to solution chemistry. Depending on the basicity of the γ-Al_2O_3 surface, which is regulated by the temperature of treatment to control the degree of dehydroxylation of the initially hydroxylated γ-Al_2O_3, the surface species formed by carbonylation of the adsorbed species formed from $[Ir(CO)_2(acac)]$ may be predominantly any of the following: $[Ir_4(CO)_{12}]$, $[HIr_4(CO)_{11}]^-$, $[Ir_8(CO)_{22}]^{2-}$, or $[Ir_6(CO)_{15}]^{2-}$ (Zhao and Gates, 1997).

In all these syntheses, the surface chemistry is analogous to known solution chemistry (Gates, 1994). Thus, syntheses on nearly neutral surfaces, for example, that of $[Ir_4(CO)_{12}]$ on γ-Al_2O_3 (Kawi, Chang, and Gates, 1993) and that of $[H_4Ru_4(CO)_{12}]$ and of $[H_4Os_4(CO)_{12}]$ on SiO_2 (Dossi *et al.*, 1990), take place as in neutral solvents. The syntheses on the basic surface of MgO take place as in basic solutions. Roberto and colleagues (1996) made SiO_2 surfaces basic by incorporation of alkali metal carbonates and tuned the basicity and thus the reactivity by varying the type and amount of the carbonate.

Sometimes solution reactions take place in the presence of reducing agents, and evidently the reducing sites on MgO play an analogous role (Lamb, Gates, and Knözinger, 1988). The reactivities of all the metal oxides are influenced by the treatment conditions and methods of preparation, which influence, for example, the degree of hydroxylation and the base strength of the surface.

The syntheses that take place on metal oxide surfaces also take place in zeolite cages. Reactions in the nearly neutral zeolite NaY are similar to those occurring on some γ-Al_2O_3 surfaces, and those in basic zeolite NaX resemble those on basic metal oxide surfaces, such as that of hydroxylated MgO. "Ship-in-a-bottle" syntheses take place when clusters formed in zeolite cages are trapped there because they are too large to fit through the apertures connecting the cages (Kawi and Gates, 1994).

Supported nanoclusters have been prepared by decarbonylation of neutral or anionic metal carbonyl clusters on supports. The decarbonylation chemistry is not fully understood. The chemistry accompanying removal of the CO ligands from metal carbonyl clusters on metal oxides evidently involves hydroxyl groups or water on the surface of the metal oxide.

For example, the decarbonylation of $[Ir_4(CO)_{12}]$ on γ-Al_2O_3 in He at 473 K takes place with little or no disruption of the tetrahedral Ir_4 frame (Kawi, Chang, and Gates, 1993). Decarbonylation of MgO-supported $[HIr_4(CO)_{11}]^-$ and of $[Ir_6(CO)_{15}]^{2-}$ takes place with near retention of the respective tetrahedral and octahedral metal frames in He at 573 K (van Zon *et al.,* 1993). However, decarbonylation of tetrairidium carbonyl clusters formed from $[Ir_4(CO)_{12}]$ on highly hydroxylated MgO takes place with sintering of the iridium to give larger, nonuniform nanoclusters or particles (Triantafillou and Gates, 1994). Sintering is affected in a complex way by surface OH groups, and generalizations about the effects of these groups seem not yet to be warranted.

B. Structural Characterization

The methods of structure determination of supported nanoclusters are essentially the same as those mentioned previously for supported metal complexes. EXAFS spectroscopy plays a more dominant role for the metal clusters than for the complexes because it provides good evidence of metal–metal bonds. Combined with density functional theory, EXAFS spectroscopy has provided much of the structural foundation for investigation of supported metal clusters. EXAFS spectroscopy provides accurate determinations of metal–metal distances ($\pm$1–2%), but it gives only average structural information and relatively imprecise values of coordination numbers. EXAFS spectroscopy provides structure data that are most precise when the clusters are extremely small (containing about six or fewer atoms) and nearly uniform (Alexeev and Gates, 2000).

C. Examples

$[Rh_6(CO)_{16}]$ was prepared on the surface of TiO_2 powder by adsorption of $[Rh(CO)_2(acac)]$ followed by treatment in CO. IR and EXAFS data indicate the presence of $[Rh_6(CO)_{16}]$ (Goellner and Gates, 2001). The decarbonylated sample is well approximated as Rh_6 on TiO_2, as shown by the EXAFS result indicating a Rh–Rh first-shell coordination number of about 4, the value for an octahedral metal frame (as in $[Rh_6(CO)_{16})$ (Goellner and Gates, 2001).

Similar results have been obtained for $[Ir_4(CO)_{12}]$ (and Ir_4 formed from it) on various supports and for $[Ir_6(CO)_{16}]$ (and Ir_6 formed from it) on various supports; details are given elsewhere (Gates, 1998). One of the most thoroughly investigated supported clusters is Ir_4 in zeolite NaX, which has been characterized by EXAFS spectroscopy and density functional theory (Ferrari *et al.,* 1999). The EXAFS data indicate an Ir–Ir coordination number of nearly 3, as expected for an Ir_4 tetrahedron, and the theory confirms the stability of Ir_4 tetrahedra assumed to be placed at six-rings in the faujasite zeolite (Fig. 3). EXAFS spectroscopy also characterizes the metal–support interface in terms of M–O coordination numbers and distances (although the uncertainty in the parameters is markedly greater than for first-shell metal–metal contributions). Both relatively short metal–support oxygen ($M–O_s$) distances of about 2.1–2.2 Å and relatively long $M–O_l$ distances of about 2.5–2.7 Å are typical EXAFS results for the metal–support interface (Koningsberger and Gates, 1992), being determined with precisions of 1 to 2%. The $M–O_s$ coordination numbers are much less precise than the distances and do not lead to strong generalizations, except that an $M–O_s$ coordination number of roughly 1 to 2 is typical of noble metal nanoclusters on

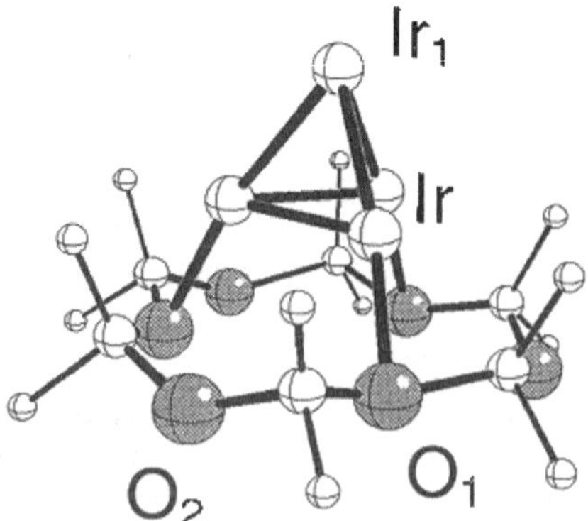

FIG. 3. Model of Ir_4 bonded at a six-ring in zeolite NaX with an Al : Si ratio of 1, as modeled by density functional theory (Ferrari *et al.,* 1999).

oxide and zeolite supports. (The quality of the EXAFS data characterizing the metal–support interactions in supported metal clusters is generally not as good as that characterizing the metal–support interactions in supported metal complexes, because the metal–support contributions are greater for the latter samples; the metal–metal contributions dominate for supported metal clusters.)

The theoretical parameters characterizing Ir_4 in zeolite NaX (Fig. 3) indicate Ir–O distances of about 2.2 Å, in good agreement with EXAFS data (Ferrari *et al.,* 1999) and approximately equal to the metal–oxygen bond distances found experimentally and theoretically for supported metal complexes, as discussed above. When the structure of Fig. 3 is rotated 60°, the theory indicates an Ir–O distance of about 2.7 Å, in agreement with the longer distances observed by EXAFS spectroscopy (but this agreement may be fortuitous).

Similarly, theoretical results for Os_5C on MgO (Fig. 4), prepared from $[Os_5C(CO)_{14}]^{2-}$, are in good agreement with EXAFS data characterizing the metal frame, and they indicate Os–O_s distances of about 2.1 Å, about the same as the values mentioned above and in good agreement with the EXAFS data. The bond distances suggest that the oxygen of the support serves as a ligand to a supported metal cluster, much as it serves as a ligand to a supported metal complex.

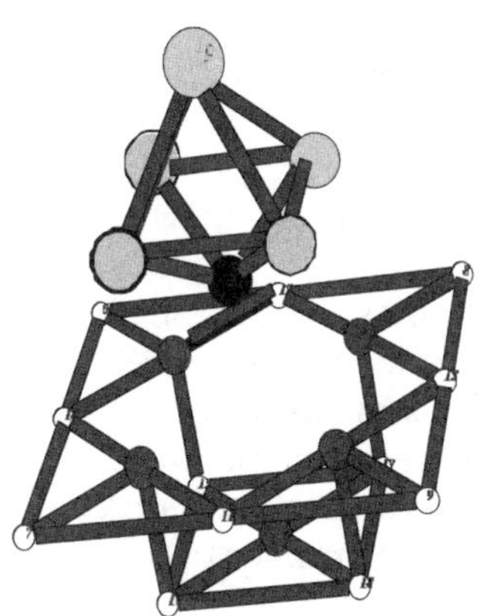

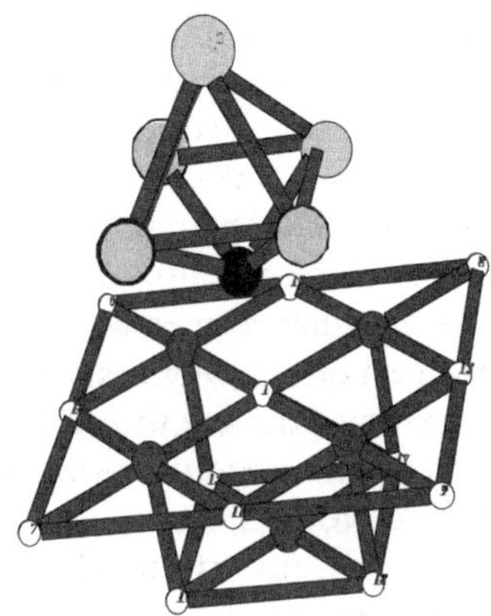

FIG. 4. Models of Os_5C on the (100) face of MgO, as modeled by density functional theory (Goellner, Neyman, *et al.,* 2000). The structure at the right is represented on the defect-free surface, and the structure at the left is represented at a cation defect site of the surface. The binding energy of the Os_5C cluster to the fragment of MgO in the structure shown at the left (where an Mg cation is missing) was calculated to be 4.89 eV, and the binding energy of the Os_5C cluster to the defect-free MgO fragment shown at the right was calculated to be 0.59 eV; thus, the former structure is much more tightly bound than the former.

When the structure at the right in Fig. 4 is rotated 90°, the distance indicated by theory is about 2.6 Å, again in good agreement with EXAFS data for the longer Os–O_l distance (Panjabi, 2000; Koningsberger and Gates, 1992) (but this agreement may be fortuitous).

The theoretical metal–metal distances in both the supported Ir_4 and Os_5C clusters are significantly (about 0.2 Å) less than the metal–metal distances determined by EXAFS spectroscopy; the most likely explanation of this difference (confirmed by theory) is that some residual ligands, such as carbon, remained on the metal frames after decarbonylation (Ferrari *et al.,* 1999; Goellner, Neyman *et al.,* 2000). Recent EXAFS results for Rh_6 on TiO_2 (Goellner and Gates, 2001) are consistent with the presence of carbon on some decarbonylated clusters—but not on others, depending on the preparation conditions.

The theoretical investigations of supported nanoclusters provide additional insights about the role of the support. Theory indicates that Ir_4 clusters on zeolite NaX are neutral or slightly negatively charged and that cluster–support bonding induces a polarization of the cluster that could affect reactivity and catalysis (Ferrari *et al.,* 1999). The theoretical investigation of supported Os_5C clusters provides evidence that the clusters on the defect-free (001) face of MgO (Fig. 4) are much less strongly bound than the clusters at cation vacancy sites, which are common on MgO; the clusters are tightly bound at the defect sites (Goellner, Neyman *et al.,* 2000). The result is suggested to be valid with some generality for nanoclusters on oxide supports; this suggestion is important for its implications regarding the stability of supported nanoclusters, for example, during catalysis.

D. Catalytic Properties

EXAFS spectra representing catalyst–support combinations, $Ir_4/\gamma\text{-}Al_2O_3$, $Ir_6/\gamma\text{-}Al_2O_3$, and Ir_4/MgO, show that the cluster frames were maintained before, during (Table IV), and after catalysis of propene hydrogenation, provided that the conditions were mild (e.g., room temperature and 1 atm) (Panjabi, Argo, and Gates, 1999), but when the temperature of catalysis reached about 150°C, the metals aggregated on the support. The data suggest that the supported clusters themselves are catalytically active. The ligands on the clusters during catalysis are not yet fully determined, and it is possible that residual ligands (such as C) remaining from the CO of the precursor clusters were present in addition to hydrocarbon and hydrogen observed by IR spectroscopy.

The rates of toluene hydrogenation catalyzed by Ir_4 and by Ir_6 supported on metal oxides and zeolites differ from each other, typically by factors of

TABLE IV
EXAFS RESULTS CHARACTERIZING SUPPORTED METAL CLUSTERS DURING CATALYTIC HYDROGENATION OF PROPENE FED IN AN EQUIMOLAR MIXTURE TO A FLOW REACTOR OPERATED AT STEADY STATE AT 25°C AND 1 ATM (PANJABI, ARGO, AND GATES, 1999)[a]

Catalyst	EXAFS parameters				
Metal cluster/ support	Shell	N	R(Å)	$10^3 \cdot \Delta\sigma^2$(Å^2)	ΔE_0(eV)
Ir_4/γ-Al_2O_3	Ir–Ir	3.2	2.71	2.4	–3.2
	Ir–$O_{support}$				
	Ir–O_s	1.3	2.17	3.5	–15.6
	Ir–O_l	0.5	2.70	–6.8	–10.9
	Ir–Al	0.3	1.77	9.3	2.6
Ir_6/γ-Al_2O_3	Ir–Ir	4.0	2.71	–5.3	–3.2
	Ir–$O_{support}$				
	Ir–O_s	1.3	2.23	1.2	–18.3
	Ir–O_l	0.5	2.72	5.2	–13.0
	Ir–Al	0.1	1.51	0.6	–3.7
Ir_4/MgO	Ir–Ir	3.0	2.72	1.5	–2.5
	Ir–$O_{support}$				
	Ir–O_s	1.3	2.15	2.0	–12.7
	Ir–O_l	0.5	2.69	–6.3	–9.9

[a] Notation: N, coordination number; R, distance between absorber and backscatterer atom; $\Delta\sigma^2$, Debye–Waller factor; ΔE_0, inner potential correction; O^* = carbonyl oxygen; the subscripts s and l refer to short and long, respectively.

several; thus, these supports are roughly equivalent, possibly because they all offer similar ligands to the metal cluster (Gates, 1998).

In prospect, structurally well-defined supported metal clusters provide the opportunity for resolving support from cluster-size effects. A family of supported Ir clusters and particles was prepared from $[Ir_4(CO)_{12}]$ on γ-Al_2O_3 (Xiao *et al.*, 1996). The smallest clusters were approximately Ir_4, and samples with larger nanoclusters and particles were prepared by treating Ir_4/γ-Al_2O_3 in H_2 under various conditions to cause aggregation and vary the average cluster or particle size. The catalytic activity was measured for each sample. The rate per exposed Ir atom increased by two orders of magnitude as the cluster or particle size increased, becoming independent of particle size when the average particle contained about 100 atoms (Xiao *et al.*, 1996). The data for the larger particles conform to the expected pattern for the structure-insensitive reaction, but those for the smaller clusters and particles do not. (A structure-insensitive reaction is one that takes place at approximately the same rate per exposed metal atom, independent of the face of the metal crystal that is exposed or the average size of the metal nanoparticle in the catalyst (Boudart and Djéga-Mariadassou, 1984)).

The cluster-size dependence has not yet been explained; it may reflect an intrinsically low activity of the clusters, but it might also be a consequence of increasing removal of residual ligands such as C from the clusters with increasing severity of treatment in H_2. Other possibilities include a steric effect of the support, limiting adsorption of the reactants on the metal—such an effect would be greatest for the smallest clusters. Electronic effects should not be ruled out.

The result that the catalytic activities of extremely small clusters are different from those of larger nanoclusters and nanoparticles, even for structure-insensitive reactions, is broadly borne out by the work of Heiz and Colleagues (1999) for catalysts made by deposition of size-selected gas-phase clusters. The results are qualitatively consistent with results showing unique reactivities of gas-phase clusters (for example, Schnabel and Irion (1992) observed that Fe_n^+ clusters ($4 \leq n \leq 13$) are unreactive for bonding and dehydrogenation of ethane, except for $n = 4$ or 5), although these results cannot be extrapolated easily to supported clusters because the support exerts a strong influence on the reactivity. Much work remains to be done to connect the chemistry of metal complexes and clusters on supports with that of metal atoms and clusters in the gas phase (Armentrout, 1999), and research with well-defined supported metal complexes and nanoclusters should help forge the link.

The supported nanoclusters described above have strong connections to industrial catalysts, although the nanoclusters in industrial catalysts are much less uniform and less susceptible to incisive structural characterization than the materials considered here. Clusters of only a few metal atoms are applied in technology, e.g., Pt clusters in zeolite LTL for alkane dehydrocyclization (Jentoft *et al.,* 1998; Vaarkamp *et al.,* 1993; Miller *et al.,* 1993). EXAFS and transmission electron microscopy indicate Pt clusters containing from 5 to 12 atoms each, on average (Jentoft *et al.,* 1998; Vaarkamp *et al.,* 1993; Miller *et al.,* 1993). PdPt clusters of approximately this size are also present in new zeolite-supported catalysts for aromatic hydroprocessing in the presence of sulfur to meet stringent new fuel quality specifications; the extremely small clusters are resistant to poisoning by sulfur (Yasuda, Sato, and Yoshimura, 1999; Baird *et al.,* 1997). Furthermore, two- and three-atom clusters of Pt on γ-Al_2O_3 (Pt/γ-Al_2O_3) were observed, along with larger Pt nanoclusters, in conventionally prepared Pt/γ-Al_2O_3 by scanning transmission electron microscopy (Nellest and Pennycook, 1996). The roles of extremely small (and possibly undetected) clusters in many practical catalysts remain to be determined; for example, they could be important in supported gold catalysts, the remarkable catalytic properties of which are not yet well understood (Haruta, 1997).

E. Generalizations about Structure, Bonding, Reactivity, and Catalysis

The results just summarized lead to the following generalizations about supported metal nanoclusters:

- There is good agreement between EXAFS data and theoretical results characterizing structure and bonding in supported metal nanoclusters and the metal–support interface.
- The results indicate bonding interactions between the clusters and oxygen atoms of the supports; the supports are regarded as polydentate ligands for the clusters, much as they are ligands for metal complexes.
- Theory indicates that supported nanoclusters typified by Ir_4 on zeolite NaX are nearly neutral.
- Theory indicates that supported nanoclusters typified by Os_5C on MgO are bonded more strongly at surface defect sites than at defect-free sites.
- Supported clusters of only a few metal atoms have new catalytic properties, different from those of larger supported nanoclusters and nanoparticles.

V. Supported Metal Nanoparticles

Most supported metal catalysts incorporate metals in nanoclusters or nanoparticles larger than those considered above. A typical nanoparticle diameter is in the range 1 to 10 nm. Metal particles in this size range are usually large enough to have bulklike properties, and it is common to refer to them as metallic. They may expose various crystal faces, which may have different catalytic activities. Catalytic properties of nanoparticles, especially those with diameters in the approximate range 1 to 5 nm, may depend strongly on the particle sizes and shapes—and their interactions with the support, which influence the size and shape—at least in part because these influence the distribution of crystal faces that are exposed (Boudart, 1985). Nanoparticles with diameters greater than about 5 nm may have fully developed crystallographic habits, so that the particle size barely affects the catalytic activity.

The properties of nanoparticles may often be significantly different from those of the smaller nanoclusters discussed above, which may have unique catalytic sites, different from those on nanoparticles (Argo and Gates, forthcoming).

ACKNOWLEDGMENTS

This work was supported by the Department of Energy, Office of Energy Research, Office of Basic Energy Sciences, a gift from Ford Motor Co., and the National Science Foundation (Grant CTS-9615257).

REFERENCES

Alexeev, O., and Gates, B. C., EXAFS characterization of supported metal-complex and metal-cluster catalysts made from organometallic precursors, *Top. Catal.* **10,** 273 (2000).

Armentrout, P. B., Gas phase organometallic chemistry, *in* "Topics in Organometallic Chemistry" (J. M. Brown and P. Hofmann, Eds.), Vol. 4, p. 1. Springer Verlag, Berlin, 1999.

Baird, W. C. Jr., McVicker, G. B., Schorfheide, J. J., Klein, D. P., Hantzer, S., Daage, M., Touvelle, M. S., Ellis, E. S., Vaughan, D. E. W., and Chen, J. G., U.S. Patent 5935420 (1997).

Basset, J.-M., Lefebvre, F., and Santini, C., Surface organometallic chemistry: Some fundamental features including the coordination effects of the support. *Coord. Chem. Rev.* **180,** 1703, (1998).

Boudart, M., Heterogeneous catalysis by metals. *J. Mol. Catal.* **30,** 27 (1985).

Boudart, M., and Djéga-Mariadassou, G., "Kinetics of Heterogeneous Catalytic Reactions." Princeton University Press, Princeton, 1984.

Brenner, A., Comparative catalytic activity of supported clusters, *in* "Metal Clusters" (M. Moskovits, Ed.), p. 249. Wiley-Interscience, New York, 1986.

Budge, J. R., Scott, J. P., and Gates, B. C., Synthesis and characterization of an alumina-supported $RuOs_3$ cluster catalyst. *J. Chem. Soc., Chem. Commun.* **1983,** 342 (1983).

Che, M., From unit operations to elementary processes: A molecular and multidisciplinary approach to catalyst preparation. *Stud. Surf. Sci. Catal.* **130,** 115 (2000).

Che, M., Clause, O., and Marcilly, C., Supported catalysts. Deposition of active component. Impregnation and ion exchange, *in* "Preparation of Solid Catalysts" (G. Ertl, H. Knözinger, and J. Weitkamp, Eds.), p. 315 Wiley-VCH, Weinheim, 1999.

Deutsch, S. E., Chang, J.-R., and Gates, B. C., Osmium subcarbonyls on γ-alumina: Characterization of the metal-support bonding by infrared, nuclear magnetic resonance, and X-ray absorption spectroscopies. *Langmuir* **9,** 1284 (1993).

Dossi, C., Psaro, R., Roberto, D., Ugo, R., and Zanderighi, G. M., Surface-mediated synthesis of $H_4Ru_4(CO)_{12}$, $H_4Os_4(CO)_{12}$ on hydroxylated silica. *Inorg. Chem.* **29,** 4368 (1990).

Ferrari, A. M., Neyman, K. M., Mayer, M., Staufer, M., Gates, B. C., and Rösch, N., Faujasite-supported Ir_4 clusters: A density functional model study of metal–zeolite interactions. *J. Phys. Chem. B* **103,** 5311 (1999).

Gates, B. C., Surface-mediated synthesis of metal clusters. *J. Mol. Catal.* **86,** 95 (1994).

Gates, B. C., Supported metal clusters: Synthesis, structure, and catalysis. *Chem. Rev.* **95,** 511–522 (1995).

Gates, B. C., Metal cluster catalysts dispersed on solid supports, *in* "Catalysis by Di- and Polynuclear Metal Cluster Complexes" (R. D. Adams and F. A. Cotton, Eds.), p. 509. Wiley-VCH, New York, 1998.

Gates, B. C., Guczi, L., and Knözinger, H., (Eds.), "Metal Clusters in Catalysis." Elsevier, Amsterdam, 1986.

Gates, B. C., and Lamb, H. H., Supported metals and supported organometallics, *J. Mol. Catal.* **52,** 1 (1989).

Goellner, J. P, and Gates, B. C., Synthesis and characterization of site-isolated hexarhodium clusters on titania powder. *J. Phys. Chem. B.* **105,** 3269 (2001).

Goellner, J. F., Gates, B. C., Vayssilov, G., and Rösch, N., Structure and bonding of a site-isolated transition metal complex: Rhodium dicarbonyl in highly dealuminated zeolite Y. *J. Am. Chem. Soc.* **122,** 8056 (2000).

Goellner, J. F., Neyman, K. M., Mayer, M., Nörtemann, F., Gates, B. C., and Rösch, N., Ligand-free osmium clusters supported on MgO. A density functional study. *Langmuir* **16,** 2736 (2000).

Haruta, M., Size-and support-dependency in the catalysis of gold. *Catal. Today* **36,** 153 (1997).

Heiz, U., Sanchez, A., Abbet, S., and Schneider, W.-D., Catalytic oxidation of carbon monoxide on monodispersed platinum clusters: Each atom counts. *J. Am. Chem. Soc.* **121,** 3214 (1999).

Heiz, U., and Schneider, W.-D., Nanoassembled model catalysts. *J. Phys. D—Appl. Phys.* **33,** R85 (2000).

Hu, A., Neyman, K. M., Staufer, M., Belling, T., Gates, B. C., and Rösch, N., A surface site as polydentate ligand of a metal complex: Density functional studies of rhenium subcarbonyls supported on magnesium oxide. *J. Am. Chem. Soc.* **121,** 4522 (1999).

Ichikawa, M., Metal cluster compounds as molecular precursors for tailored metal catalysts. *Adv. Catal.* **38,** 283 (1992).

Iwasawa, Y., Chemical design surfaces for active solid catalysts. *Adv. Catal.* **35,** 187 (1987).

Iwasawa, Y., Chemical design and characterization of catalysts and catalysis—an approach to dynamic catalyst design. *Catal. Today* **18,** 21–72 (1993).

Jentoft, R. E., Tsapatsis, M. A., Davis, M. E., and Gates, B. C., Platinum clusters supported in zeolite LTL: Influence of catalyst morphology on performance in *n*-hexane reforming. *J. Catal.* **179,** 565 (1998).

Kaminsky, W., Metallocene catalysts for olefin polymerization, *in* "Science and Technology in Catalysis 1998" (V. Dragutan and R. Streck, Eds.), p. 3. Elsevier, Amsterdam, 1999.

Kawi, S., Chang, J.-R., and Gates, B. C., Tetrairidium clusters supported on γ-Al_2O_3: Formation from $[Ir_4(CO)_{12}]$ and CO-induced morphology changes. *J. Phys. Chem.* **97,** 5375 (1993).

Kawi, S., and Gates, B. C., Clusters in cages, *in* "Clusters and Colloids" (G. Schmid, Ed.), p. 299. VCH, Weinheim, 1994.

Kirlin, P. S., DeThomas, F. A., Bailey, J. W., Gold, H. S., Dybowski, C., and Gates, B. C., Molecular oxide-supported rhenium carbonyl complexes: Synthesis and characterization by vibrational spectroscopy. *J. Phys. Chem.* **90,** 4882 (1986).

Kirlin, P. S., and Gates, B. C., Activation of the C–C bond provides a molecular basis for structure sensitivity in catalysis. *Nature* (*London*) **325,** 38 (1987).

Kirlin, P. S., Knözinger, H., and Gates, B. C., Mononuclear, trinuclear, and metallic rhenium catalysts supported on MgO: Effects of structure on catalyst performance. *J. Phys. Chem.* **94,** 8451 (1990).

Kirlin, P. S., van Zon, F. B. M., Koningsberger, D. C., and Gates, B. C., Surface catalytic sites prepared from $[HRe(CO)_5]$ and $[H_3Re_3(CO)_{12}]$: Mononuclear, trinuclear, and metallic rhenium catalysts supported on MgO. *J. Phys. Chem.* **94,** 8439 (1990).

Koningsberger, D. C., and Gates, B. C., Nature of the metal–support interface in supported metal catalysts: Results from X-ray absorption spectroscopy. *Catal. Lett.* **14,** 271 (1992).

Kristen, M. O., Supported metallocene catalysts with MAO and boron activators. *Top. Catal.* **7,** 89 (1999).

Lamb, H. H., Gates, B. C., and Knözinger, H., Molecular organometallic chemistry on surfaces: Reactivity of metal carbonyls on metal oxides. *Angew. Chem. Int. Ed. Engl.* **27,** 1127 (1988).

Li, X.-J., Gates, B. C., Knözinger, H., and Alizo Delgado, E., Support effects in hexene-1 isomerization catalyzed by oxide-supported osmium complexes. *J. Catal.* **88,** 355 (1984).

Meiwes-Broer, K.-H., (Ed.), "Metal Clusters at Surfaces." Springer, Heidelberg, 2000.

Messerli, S., Schintke, S., Morgenstern, K., Sanchez, A., Heiz, U., and Schneider, W.-D., Imaging size-selected silicon clusters with a low-temperature scanning tunneling microscope. *Surf. Sci.* **465,** 331 (2000).

Miessner, H., Surface-chemistry in a zeolite matrix—Well-defined dinitrogen complexes of rhodium supported on dealuminated Y-zeolite. *J. Am. Chem. Soc.* **116,** 11522 (1994).

Miessner, H., Burkhardt, I., Gutschick, D., Zecchina, A., Morterra, C., and Spoto, G., The formation of a well-defined rhodium dicarbonyl in highly dealuminated rhodium-exchanged zeolite-Y by interaction with CO. *J. Chem. Soc., Faraday Trans. I* **85,** 2113 (1989).

Miller, J. T., Meyers, B. L., Modica, F. S., Lane, G. S., Vaarkamp, M., and Koningsberger, D. C., Hydrogen temperature-programmed desorption (H_2 TPD) of supported platinum catalysts. *J. Catal.* **143,** 395 (1993).

Nellist, P. D., and Pennycook, S. J., Direct imaging of the atomic configuration of ultradispersed catalysts. *Science* **274,** 413 (1996).

Panjabi, G., Argo, A. M., and Gates, B. C., Supported iridium cluster catalysts for propene hydrogenation: Identification by X-ray absorption spectra measured during catalysis. *Chem. Eur. J.* **5,** 2417 (1999).

Panjabi, G. A., Salvi, S. N., Phillips, B. L., Allard, L. F., and Gates, B. C., Ph.D. Thesis, University of California, Davis (2000).

Papile, C. J., and Gates, B. C., Rhenium subcarbonyls on magnesium oxide: Identification of the surface oxo and hydroxo ligands by infrared spectroscopy. *Langmuir* **8,** 74 (1992).

Purnell, S. K., Xu, X., Goodman, D. W., and Gates, B. C., Adsorption and reaction of $[Re_2(CO)_{10}]$ on ultrathin MgO films grown on a Mo(110) surface: Characterization by infrared reflection–absorption spectroscopy and temperature-programmed desorption. *J. Phys. Chem.* **98,** 4076 (1994a).

Purnell, S. K., Xu, X., Goodman, D. W., and Gates, B. C., Adsorption and reaction of $[HRe(CO)_5]$ on ultrathin MgO(111) and amorphous SiO_2 films grown on a Mo(110) Surface: Characterization by infrared reflection–absorption spectroscopy and temperature-programmed desorption. *Langmuir* **10,** 3057 (1994b).

Roberto, D., Cariati, E., Ugo, R., and Psaro, R., Surface-mediated organometallic synthesis: High-yield preparations of neutral and anionic osmium carbonyl clusters by controlled reduction of silica-supported $[Os(CO)_3Cl_2]_2$ and $OsCl_3$ in the presence of Na_2CO_3 or K_2CO_3. *Inorg. Chem.* **35,** 2311 (1996).

Roscoe, S. B., Frechet, J. M. J., Walzer, J. F., and Dias, A. J., Polyolefin spheres from metallocenes supported on noninteracting polystyrene. *Science* **280,** 270 (1998).

Schnabel, P., and Irion, M. F., Reactions of $Fe_n^+(3 \leq n \leq 12)$ with ethane, ethylene and acetylene—reaction paths and rate constants. *Ber. Bunsenges. Phys. Chem.* **96,** 1101 (1992).

Sullivan, D. L., Roark, R. D., Ekerdt, J. G., Deutsch, S. E., and Gates, B. C., Reaction and spectroscopic study of silica-supported molybdenum(IV) and tungsten(IV) dimers. *J. Phys. Chem.* **99,** 3678 (1995).

Triantafillou, N. D., and Gates, B. C., Magnesia-supported tetrairidium clusters derived from $[Ir_4(CO)_{12}]$. *J. Phys. Chem.* **98,** 8431 (1994).

Triantafillou, N. D., Purnell, S. K., Papile, C. J., Chang, J.-R., and Gates, B. C., A Family of rhenium subcarbonyls on MgO: Structural characterization by extended X-ray absorption fine structure spectroscopy. *Langmuir* **10,** 4077 (1994).

Vaarkamp, M., Modica, F. S., Miller, J. T., and Koningsberger, D. C., Influence of hydrogen pretreatment on the structure of the metal–support interface in Pt/zeolite catalysts. *J. Catal.* **144,** 611 (1993).

van Zon, F. B. M., Maloney, S. D., Gates, B. C., and Koningsberger, D. C., Structure and nature of the metal–support interface: Characterization of iridium clusters on magnesium oxide by extended X-ray absorption fine structure spectroscopy. *J. Am. Chem. Soc.* **115,** 10317 (1993).

van't Blik, H. F. J., van Zon, J. B. A. D., Huizinga, T., Vis, J. C., Koningsberger, D. C., and Prins, R., Structure of rhodium in an ultradispersed Rh/Al_2O_3 catalyst as studied by EXAFS and other techniques. *J. Am. Chem. Soc.* **107,** 3139 (1985).

Vidal, V., Theolier, A., Thivolle Cazat, J., and Basset, J.-M., Metathesis of alkanes catalyzed by silica-supported transition metal hydrides. *Science* **276,** 99 (1997).

Wovchko, E. A., and Yates, J. T., Chemical bond activation on surface sites generated photochemically from $Rh(I)(CO)_2$ species. *Langmuir* **15,** 3506 (1999).

Xiao, F.-S., Weber, W. A., Alexeev, O., and Gates, B. C., Probing the limits of structure insensitivity: Size-dependent catalytic activity of Al_2O_3-supported iridium clusters and particles for toluene hydrogenation. *Stud Surf. Sci. Catal.* **101,** 1135 (1996).

Xu, Z., Xiao, F.-S., Purnell, S. K., Alexeev, O., Kawi, S., Deutsch, S. E., and Gates, B. C., Size-dependent catalytic activity of supported metal clusters. *Nature* (*London*) **372,** 346 (1994).

Yasuda, H., Sato, T., and Yoshimura, Y., Influence of the acidity of USY zeolite on the sulfur tolerance of Pd-Pt catalysts for aromatic hydrogenation. *Catal. Today* **50,** 63 (1999).

Zhao, A., and Gates, B. C., Probing metal oxide surface reactivity with adsorbate organometallic chemistry: Formation of iridium carbonyl clusters on γ-Al_2O_3, *Langmuir* **13,** 4024 (1997).

NANOSTRUCTURED ADSORBENTS

Ralph T. Yang

Department of Chemical Engineering, University of Michigan
Ann Arbor, Michigan 48109

I.	Introduction	80
II.	Fundamental Factors for Designing Adsorbents	81
	A. Potential Energies for Adsorption	81
	B. Heat of Adsorption	83
	C. Effects of Adsorbate Properties on Adsorption: Polarizability (α), Dipole Moment (μ), and Quadrupole Moment (Q)	84
	D. Basic Considerations for Sorbent Design	85
III.	Activated Carbon, Activated Alumina, and Silica Gel	88
	A. Recent Developments on Activated Carbon	91
	B. Activated Alumina and Silica Gel	93
IV.	MCM-41	94
V.	Zeolites	96
	A. Structures and Cation Sites	98
	B. Unique Adsorption Properties: Anionic Oxygens and Isolated Cations	99
	C. Interactions with Cations: Effects of Site, Charge, and Ionic Radius	100
VI.	π-Complexation Sorbents	108
	A. π-Complexation Sorbents for Olefin–Paraffin Separations	109
	B. Effects of Cation, Anion, and Substrate	112
	C. Nature of the π-Complexation Bond	114
	D. Olefin–Diene Separation and Purification, Aromatic and Aliphatics Separation, and Acetylene Separation	117
VII.	Other Sorbents and Their Unique Adsorption Properties: Carbon Nanotubes, Heteropoly Compounds, and Pillared Clays	118
	A. Carbon Nanotubes	118
	B. Heteropoly Compounds	119
	C. Pillared Clays	120
	References	121

ADVANCES IN CHEMICAL ENGINEERING, VOL. 27

0065-2377/01 $35.00

This chapter discusses the fundamental principles for designing nanoporous adsorbents and recent progress in new sorbent materials. For sorbent design, detail discussion is given on both fundamental interaction forces and the effects of pore size and geometry on adsorption. A summary discussion is made on recent progress on the following types of materials as sorbents: activated carbon, activated alumina, silica gel, MCM-41, zeolites, π-complexation sorbents, carbon nanotubes, heteropoly compounds, and pillared clays.

I. Introduction

Since the development of synthetic zeolites and pressure swing adsorption (PSA) cycles, adsorption has been playing an increasingly important role in gas separation and purification in chemical and petrochemical industries (Yang, 1997). Even though new applications for gas separations by adsorption are continually being developed, the most important applications have been air separation (for production of O_2 and N_2) and hydrogen separation (from fuel gases). Approximately 20% of the O_2 and N_2 are currently being produced by PSA, and the share is continuing to increase. Drying by PSA is already a mature technology. Other applications, including environmental control, are on the horizon.

The increasing industrial applications for adsorption have stimulated a growing interest in research. The research has been advancing on several fronts: thermodynamics of adsorption (particularly statistical mechanics), diffusion of pores, PSA simulation, new process and cycle development, sorbent characterization, and development of new sorbents. Significant advances have been made on all fronts during the last decade.

Applications for adsorption have been limited by the availability of sorbents. The economics of all adsorption processes are also limited by the sorbent; improvement in the sorbent will lead to improved economics. Hence, major advances in gas adsorption technology will come from the development of new sorbents. For this reason, special attention will be given to new sorbent development during the last decade.

The adsorptive separation is achieved by one of the three mechanisms: steric, kinetic, or equilibrium effect. The steric effect derives from the molecular sieving property of zeolites. In this case only small and properly shaped molecules can diffuse into the adsorbent, whereas other molecules are totally excluded. Kinetic separation is achieved by virtue of the differences in diffusion rates of different molecules. A large majority of processes operate through the equilibrium adsorption of mixture and hence are called *equilibrium separation processes.*

There are only four types of sorbents that have dominated the commercial use of adsorption: activated carbon, molecular-sieve zeolites, silica gel, and activated alumina. Estimates of worldwide annual sales of these sorbents are as follows (Humphry and Keller, 1997):

activated carbon	\$1 billion;
molecular-sieve zeolites	\$100 million;
silica gel	\$27 million;
activated alumina	\$26 million.

Among these sorbents, only activated carbon is *hydrophobic.* However, water vapor also adsorbs, and it does decrease the sorbent capacity for hydrocarbons quite substantially (Doong and Yang, 1987; Huggahalli and Fair, 1996; Russel and LeVan, 1997).

The availability of new sorbents is limiting the application of adsorption for new applications in separation and purification processes.

II. Fundamental Factors for Designing Adsorbents

Selection or synthesis of adsorbents for a given target adsorbate molecule is based on the adsorption isotherm. With the availability of high-speed computing, it is now possible to calculate the adsorption isotherms based on: (1) interaction potentials and (2) structure or geometry of the adsorbent. Hence we begin with a review of the basic forces between the adsorbent and adsorbate, paying particular attention to adsorbent design.

A. Potential Energies for Adsorption

Adsorption occurs when the interaction potential energy ϕ is equal to the work done to bring a gas molecule to the adsorbed state. As a first approximation, the adsorbed state is assumed to be at the saturated vapor pressure.

$$-\phi = -\Delta G = \int_{P}^{P_0} V\,dP = RT \ln \frac{P_0}{P}, \tag{1}$$

where ΔG is the free energy change and P_0 is the saturated vapor pressure. Hence P is the pressure when adsorption occurs for the given ϕ. (So ϕ is actually the sorbate–sorbate interaction energy on the liquid surface.)

The total potential between the adsorbate molecules and the adsorbent is the sum of the total adsorbate–adsorbate and the adsorbate–adsorbent potentials:

$$\phi_{\text{total}} = \phi_{\text{adsorbate-adsorbate}} + \phi_{\text{adsorbate-adsorbent}} \tag{2}$$

The adsorbent has only a secondary effect on the adsorbate–adsorbate interaction. For this reason, we will focus our attention on the second term, adsorbate–adsorbent potential, and refer to this term as ϕ.

There are three basic types of contributions to the adsorbate–adsorbent interactions: dispersion, electrostatic, and chemical bond. The latter, chemical bond, has been explored for adsorption only recently. Weak chemical bonds, particularly the broad type of bonds involving π electrons, or π-complexation, offer promising possibilities for designing new and highly selective sorbents. The subject of π-complexation sorbents will be discussed in a separate section. For physical adsorption, the adsorbate–adsorbent potential is

$$\phi = \phi_D + \phi_R + \phi_{\mathrm{Ind}} + \phi_{F\mu} + \phi_{\dot{F}Q}, \tag{3}$$

where ϕ_D = dispersion energy, ϕ_R = close-range repulsion energy, ϕ_{Ind} = induction energy (interaction between an electric field and an induced dipole), $\phi_{F\mu}$ = interaction between an electric field (F) and a permanent dipole (μ), and $\phi_{\dot{F}Q}$ = interaction between the field gradient and a quadrupole (Q).

The first two contributions ($\phi_D + \phi_R$) are *nonspecific* (Barrer, 1978), which are operative in all sorbate–sorbent systems. The last three contributions arise from charges (which create electric fields) on the solid surface. (This is a simplified view, because an adsorbate molecule with a permanent dipole can also induce a dipole in the sorbent if the sorbent is a conductor (Masel, 1996)). For activated carbon, the nonspecific interactions dominate. For metal oxides and ionic solids, the electrostatic interactions often dominate, depending on the adsorbate. For adsorbate with a quadrupole, the net interaction between a uniform field and the quadrupole is zero. However, the quadrupole interacts strongly with the field gradient, hence the term $\phi_{\dot{F}Q}$.

The individual contributions to the total potential have been reviewed and discussed in detail in the literature (Barrer, 1978; Masel, 1996; Razmus and Hall, 1991; Gregg and Sing, 1982; Steele, 1974; Adamson, 1976; Rigby *et al.,* 1986; Israelachilli, 1992; Young and Crowell, 1962; Ruthven, 1984; Ross and Olivier, 1964). Their functional forms are summarized here. All interactions are given between an atom (or a charge) on the surface and the adsorbate molecule.

Dispersion:

$$\phi_D = -\frac{A}{r^6}. \tag{4}$$

Repulsion:

$$\phi_R = +\frac{B}{r^{12}}. \tag{5}$$

Field (of an ion) and induced-point dipole:

$$\phi_{\text{Ind}} = -\frac{1}{2}\alpha F^2 = -\frac{\alpha q^2}{2r^4(4\pi\epsilon_0)^2}. \tag{6}$$

Field (of an ion) and point dipole:

$$\phi_{F\mu} = -F\mu\cos\theta = -\frac{q\mu\cos\theta}{r^2(4\pi\epsilon_0)}. \tag{7}$$

Field gradient ($\dot{F}$) and linear point quadrupole:

$$\phi_{\dot{F}Q} = \frac{1}{2}Q\dot{F} = -\frac{Qq(3\cos^2\theta - 1)}{4r^3(4\pi\epsilon_0)}, \tag{8}$$

where A and B are constants, α = polarizability, F = electric field, q = electronic charge of the ion on the surface, ϵ_0 = permittivity of a vacuum, μ = permanent dipole moment, θ = angle between the direction of the field or field gradient and the axis of the dipole or linear quadrupole, and Q = linear quadrupole moment (+ or –). The important parameter, r, is the distance between the centers of the interacting pair.

The dispersion and repulsion interactions form the Lennard–Jones (Barrer, 1978; Masel, 1996; Razmus and Hall, 1991; Gregg and Sing, 1982; Steele, 1974; Adamson, 1976; Rigby *et al.*, 1986) potential, with an equilibrium distance (r_0) where $\phi_D + \phi_R = 0$. This distance is taken as the mean of the van der Waals radii of the interacting pair. Once the attractive, dispersion constant, A, is known, B is readily obtained by setting at $d\phi/dr = 0$ at r_0. Hence, $B = Ar_0^6/2$. The most commonly used expression for calculating A is the Kirkwood–Muller formula:

$$A = \frac{6mc^2\alpha_i\alpha_j}{(\alpha_i/\chi_i) + (\alpha_j/\chi_j)}, \tag{9}$$

where m is the mass of an electron, c is the speed of light, χ is the magnetic susceptibility, and i and j refer to the two interacting atoms or molecules. For $\phi_{F\mu}$ and $\phi_{\dot{F}Q}$, the maximum potentials are obtained when the dipole or quadrupole are arranged linearly with the charge on the surface.

B. Heat of Adsorption

In the last section, we summarized the different contributions to the potential energy for the interactions between an adsorbate molecule (or atom) and an atom on the solid surface. To calculate the interaction energy between the adsorbate molecule and all atoms on the surface, pairwise additivity is generally assumed. The task is then to sum the interactions, pairwise, with all atoms on the surface, by integration.

It can be shown (Barrer, 1978; Ross and Olivier, 1964) that the isosteric heat of adsorption (ΔH) at low coverage is related to the sorbate–sorbent interaction potential (ϕ) by

$$\Delta H = \phi - RT + F(T), \tag{10}$$

where $F(T)$ arises due to the vibrational and translational energies of the adsorbate molecule, and for monatomic classical oscillators, $F(T) = 3RT/2$ (Barrer, 1978). For ambient temperature, $\Delta H \approx \phi$.

C. Effects of Adsorbate Properties on Adsorption: Polarizability (α), Dipole Moment (μ), and Quadrupole Moment (Q)

For a given sorbent, the sorbate–sorbent interaction potential depends on the properties of the sorbate. Among the five different types of interactions, the nonspecific interactions, ϕ_D and ϕ_R, are nonelectrostatic. The most important property that determines these interactions (and also ϕ_{Ind}) is the polarizability, α. On a surface without charges such as graphite, $\phi_{\text{Ind}} = 0$. The value of α increases with the molecular weight, because more electrons are available for polarization. From the expressions for ϕ_D, ϕ_R, and ϕ_{Ind}, it is seen that these energies are nearly proportional to α. The dispersion energy also increases with the magnetic susceptibility, χ, but not as strongly as α.

Table I gives a summary of interaction energies for a number of sorbate–sorbent pairs. Here groupings are made for the theoretical nonelectrostatic ($\phi_D + \phi_R + \phi_{\text{Ind}}$) and the electrostatic ($\phi_{F\mu} + \phi_{\dot{F}Q}$) energies.

The nonelectrostatic energies depend directly on the polarizability of the sorbate molecule. χ makes a contribution to the dispersion energy, and χ also increases with molecular weight.

Two types of sorbents are included in Table I, one without electric charges on the surface (graphitized carbon) and one with charges (zeolites). On carbon, dispersion energy dominates. On zeolites, the permanent dipole and quadrupole can make significant contributions—and, indeed, can dominate—the total energy. N_2 has a strong quadrupole but no permanent dipole; hence, $\phi_{F\mu} = 0$. From Table I shows that $\Phi_{\dot{F}Q}$ accounts for about one-third of the energies on chabazite and Na/mordenite. Na/X zeolite contains more Na^+ ions because its Si/Al ratio is lower than the other two zeolites. Consequently $\phi_{\dot{F}Q}$ contributes about one-half of the interaction energies for N_2 on Na/X. The other sorbate molecules included in Table I have both strong dipoles and quadrupoles (except H_2O, which has a strong dipole only). For adsorption of these molecules on zeolites, the ($\phi_{F\mu} + \phi_{\dot{F}Q}$) interactions clearly dominate.

A comparison of N_2 and O_2 is of particular interest for the application of air separation. Both molecules are nonpolar and have very similar

TABLE I
CONTRIBUTIONS (THEORETICAL) TO INITIAL (NEAR ZERO LOADING) HEAT OF ADSORPTION (EXPERIMENTAL, $-\Delta H$, kcal/mol) (BARRER, 1978; ROSS AND OLIVIER, 1964)

Sorbent	Sorbate[a]	$\alpha \times 10^{24}$ cm^3/molecule	$-\Delta H$	$-(\phi_D + \phi_R + \phi_{Ind})$[b]	$-(\phi_{F\mu} + \phi_{\dot{F}Q})$
Graphitized Carbon	Ne	0.396	0.74	0.73	0
	Ar	1.63	2.12	1.84	0
	Kr	2.48	2.8	2.48	0
	Xe	4.04	3.7	3.1	0
Chabazite	N_2	1.74	8.98	6.45	2.55
	N_2O	3.03	15.3	9.07	6.18
	NH_3	2.2	31.5	7.5	23.8
Na/Mordenite	N_2	1.74	7.0	4.5	2.50
	CO_2	2.91	15.7	6.73	8.93
Na/X	N_2	1.74	6.5	3.10	3.4
	CO_2	2.91	12.2	4.20	7.98
	NH_3	2.2	17.9	3.75	14.2
	H_2O	1.45	≈33.9	2.65	≈31.3

[a]Permanent dipole moments (μ, debye): N_2O = 0.161, NH_3 = 1.47, H_2O = 1.84, all others = 0. Quadrupole moments (Q, erg$^{1/2}$ cm$^{5/2}$ × 10^{26}): N_2 = –1.5, N_2O = –3.0, NH_3 = –1.0, CO_2 = –4.3, all others ≈ 0.
[b]For graphitized carbon, $\phi_{Ind} = 0$.

polarizabilities and magnetic susceptibilities. However, their quadrupole moments differ by nearly a factor of four ($Q = -0.4$ esu for O_2 and –1.5 esu for N_2). As a result, the adsorption isotherms of N_2 and O_2 on carbon are similar, whereas the isotherm of N_2 is much higher than that of O_2 on zeolites. The contribution of interaction between the field gradient and the quadrupole moment of N_2 accounts for about one-half of the total energy for N_2 adsorption on Na/X zeolite, as shown in Table I. The $\phi_{\dot{F}\mu}$ energy for O_2 is approximately one-fourth of that for N_2 (see Eq. (8)).

D. BASIC CONSIDERATIONS FOR SORBENT DESIGN

1. *Polarizability (α), Electronic Charge (q), and van der Waals Radius (r)*

For van der Waals (dispersion) interactions, the polarizabilities of the sorbate molecule and the atoms on the sorbent surface are both important (see Eq. (9)). For electrostatic interactions, for a given sorbate molecule, the charges and van der Waals radii of the surface atoms are important. The roles of these parameters are discussed separately.

TABLE II
POLARIZABILITIES (α) OF GROUND-STATE ATOMS AND IONS (IN 10^{-24} cm^3)

Atom	α	Atom	α	Atom	α
C	1.76	Li	24.3	Al	6.8
N	1.10	Na	24.08	Si	5.38
O	0.802	K	43.4	Fe	8.4
F	0.557	Rb	47.3	Co	7.5
S	2.90	Cs	59.6	Ni	6.8
Cl	2.18	Mg	10.6		
Br	3.05	Ca	22.8	Li^+	0.029
I	5.35	Sr	27.6	Na^+	0.180
		Ba	39.7	K^+	0.840
				Ca^{2+}	0.471
				Sr^{2+}	0.863
				Ba^{2+}	1.560

For a given sorbate molecule, its dispersion interaction potential with a surface atom increases with the polarizability of that surface atom. The polarizability increases with atomic weight for elements in the same family, whereas it decreases with increasing atomic weight for elements in the same row of the periodic table as the outer-shell orbitals are being increasingly filled. The polarizabilities of selected atoms are given in Table II. It is seen that the alkali and alkaline earth metal atoms have very high polarizabilities. Hence these elements, when present on the surface, can cause high dispersion potentials. When these elements are present as cations, however, the polarizabilities are drastically reduced. The polarizabilities of selected cations are also included in Table II for comparison.

For electrostatic interactions, the charges (q) and the van der Waals radii of the surface atoms (or ions) are most important. For ionic solids with point charges distributed on the surface, the positive and negative fields can partially offset when they are spaced closely. However, anions are normally bigger than cations. Consequently, the surface has a negative electric field. All electrostatic interaction potentials are proportional to q ($\Phi_{F\mu}$ and $\Phi_{\dot{F}Q}$) or q^2 (Φ_{Ind}) and inversely proportional to r^n (where $n = 2, 3, 4$; see Eqs. 6–8). Here r is the distance between the centers of the interacting pair, which should be the sum of the van der Waals radii of the two interacting atoms. Hence the van der Waals radii of the ions on the surface are important. The strong effects of charge (q) and ionic radius of the cation on the adsorption properties of ion-exchanged zeolites are discussed in a later section.

Because the ionic radius determines the distance r, it has a strong effect on the electrostatic interactions. The ionic radii of selected cations are given in Table III.

TABLE III
IONIC RADII, r_i (Å)

Ion	r_i	Ion	r_i
Li^+	0.68	Al^{3+}	0.51
Na^+	0.97	Ce^{3+}	1.03
K^+	1.33	Cu^{+1}	0.96
Rb^+	1.47	Cu^{2+}	0.72
Cs^+	1.67	Ag^+	1.26
Mg^{2+}	0.66	Ag^{2+}	0.89
Ca^{2+}	0.99	Au^{+1}	1.37
Sr^{2+}	1.12	Ni^{2+}	0.69
Ba^{2+}	1.34	Ni^{3+}	0.62

2. *Pore Size and Geometry*

The potentials discussed previously are those between two molecules or atoms. The interactions between a molecule and a flat, solid surface are greater because the molecule interacts with all adjacent atoms on the surface, and these interactions are assumed pairwise additive. When a molecule is placed between two flat surfaces, i.e., in a slit-shaped pore, it interacts with both surfaces, and the potentials on the two surfaces overlap. The extent of the overlap depends on the pore size. For cylindrical and spherical pores, the potentials are still greater because more surface atoms interact with the adsorbate molecule.

The effects of the pore size and pore geometry are best illustrated by Table IV. Table IV lists the threshold pressure for adsorption in different pore sizes and geometries for N_2 on carbon. The calculation was based on

TABLE IV
THRESHOLD PRESSURE FOR ADSORPTION IN DIFFERENT PORE SIZES AND SHAPES; N_2 ON CARBON AT 77 K; $P_0 = 1$ ATM

Pore size (Å)	P/P_0 for slit shape	P/P_0 for cylindrical shape	P/P_0 for spherical shape
3	6.7×10^{-9}	1.1×10^{-13}	5.4×10^{-52}
4	6.3×10^{-7}	1.3×10^{-12}	3.2×10^{-51}
5	9.1×10^{-6}	2.9×10^{-10}	1.1×10^{-42}
6	3.5×10^{-5}	8.3×10^{-9}	2.5×10^{-36}
7	1.2×10^{-4}	6.5×10^{-8}	6.2×10^{-32}
9	6.1×10^{-4}	3.5×10^{-6}	3.1×10^{-24}
12	2.6×10^{-3}	2.3×10^{-5}	1.2×10^{-20}
15	6.1×10^{-3}	3.2×10^{-4}	1.7×10^{-16}
20	1.4×10^{-2}	1.2×10^{-3}	6.1×10^{-13}

the Horvath–Kawazoe (HK) model (Horvath and Kawazoe, 1983), using the corrected version by Rege and Yang (2000). The corrected HK model has been shown to give pore dimensions from N_2 isotherms that agreed well with the actual pore dimension for a number of materials, including carbon and zeolites (Rege and Yang, 2000). The model is based on equating the work done for adsorption (Eq.(1)) to the total sorbate–sorbent and sorbate–sorbate interactions. The latter was the sum with all sorbate surface atoms using the Lennard–Jones potentials The results in Table IV exhibit the remarkable attraction forces acting on the adsorbate molecule due the overlapping potentials from the surrounding walls. The same carbon atom density on the surface was assumed for all geometries, i.e., 3.7×10^{15} l/cm^2. The experimental data on two molecular-sieve carbons agreed with predictions for slit-shaped pores. No experimental data are available for cylindrical pores and spherical pores of carbon. Data on these shapes may become available with the availability of carbon nanotubes and fullerenes (if an opening to the fullerene can be made).

As expected, the total interaction energies depend strongly on the van der Waals radii (of both sorbate and sorbent atoms) and the surface densities. This is true for both HK type models (Saito and Foley, 1991; Cheng and Yang, 1994) and more detailed statistical thermodynamics (or molecular simulation) approaches (such as Monte Carlo and density functional theory). Knowing the interaction potential, molecular simulation techniques enable the calculation of adsorption isotherms (see, for example, Razmus and Hall, (1991) and Cracknell *et al.* (1995)).

III. Activated Carbon, Activated Alumina, and Silica Gel

Activated carbon is the most widely used sorbent. Its manufacture and use date back to the 19th century (Jankowska, Swiatkowski, and Choma, 1991). The pore-size distribution of a typical activated carbon is given in Fig. 1, along with several other sorbents. Excellent reviews on activated carbon are available elsewhere (Jankowska, Swiatkowski, and Choma, 1991; Rouquerol, Rouquerol, and Sing, 1999; Barton *et al.,* 1999).

The raw materials for activated carbon are carbonaceous matters, such as wood, peat, coals, petroleum coke, bones, coconut shells, and fruit nuts. Anthracite and bituminous coals have been the major sources. Starting with the initial pores present in the raw material, more pores, with desired size distributions, are created by the so-called activation process. After initial treatment and pelletizing, one activation process involves carbonization at

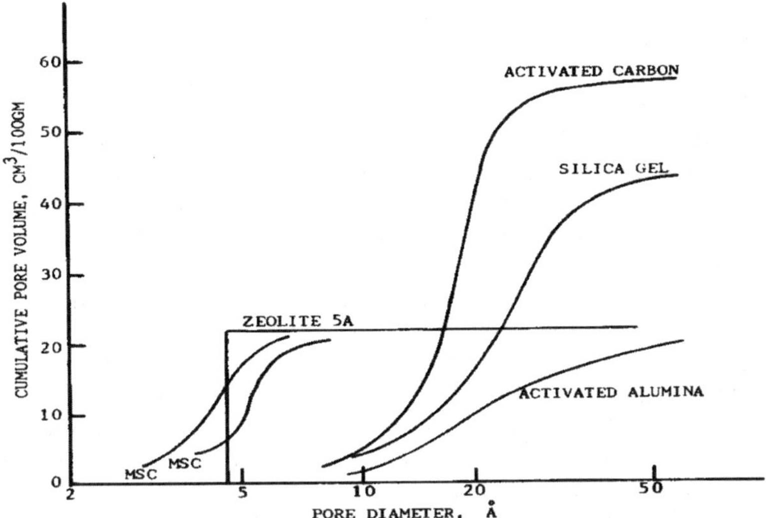

FIG. 1. Pore-size distribution for activated carbon, silica gel, activated alumina, two molecular-sieve carbons, and zeolite 5A (Yang, 1997).

400–500°C to eliminate the bulk of the volatile matter and then partial gasification at 800–1000°C to develop the porosity and surface area. A mild oxidizing gas such as CO_2 steam, or flue gas, is used in the gasification step because the intrinsic surface reaction rate is much slower than the pore diffusion rate, thereby assuring the uniform development of pores throughout the pellet. The activation process is usually carried out in fixed beds, but in recent years fluidized beds have also been used. The activated carbon created by the activation process is used primarily for the gas and vapor adsorption processes. The other activation process that is used commercially depends on the action of inorganic additives to degrade and dehydrate the cellulosic materials and, simultaneously, to prevent shrinkage during carbonization. Lignin, usually the raw material that is blended with activators such as phosphoric acid, zinc chloride, potassium sulfide, or potassium thiocyanate, is carbonized at temperatures up to 900°C. The product, usually in powder form, is used for aqueous- or gas-phase purposes. The inorganic material contained in activated carbon is measured as ash content, generally in the range between 2 and 10%.

By judicious choice of the precursor and also careful control of both carbonization and activation steps, it is possible to tailor the pore structure for particular applications (Barton *et al.*, 1999). There is now a reasonable understanding of the carbonization and activation processes (Barton *et al.*, 1999).

Mesoporosity (near or larger than 30 Å) is desirable for liquid phase applications, whereas smaller pore sizes (10 to 25 Å) are required for gas-phase applications (Yang, 1997).

The unique surface property of activated carbon, in contrast to the other major sorbents, is that its surface is nonpolar or only slightly polar as a result of the surface oxide groups and inorganic impurities. This unique property gives activated carbon the following advantages:

1. It is the only commercial sorbent used to perform separation and purification processes without requiring prior stringent moisture removal, such as is needed in air purification. (It is also useful in aqueous processes.)
2. Because of its large accessible internal surface, it adsorbs more nonpolar and weakly polar organic molecules than other sorbents do.
3. The heat of adsorption, or bond strength, is generally lower on activated carbon than on other sorbents. Consequently, stripping of the adsorbed molecules is easier and results in lower energy requirements for regeneration of the sorbent.

It is not correct, however, to regard activated carbon as hydrophobic. The equilibrium sorption of water vapor on an anthracite-derived activated carbon is compared with that of other sorbents in Fig. 2. The sorption of water vapor on activated carbon follows a Type V isotherm (according to

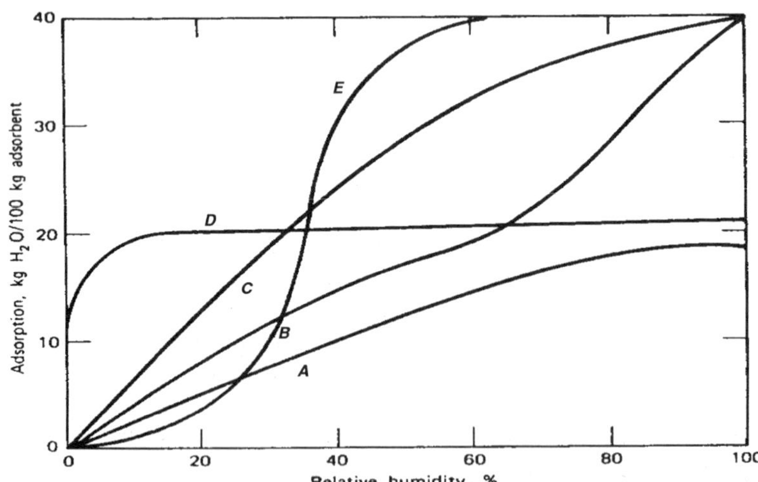

FIG. 2. Equilibrium sorption of water vapor from atmospheric air at 25°C on (*A*) alumina (granular); (*B*) alumina (spherical); (*C*) silica gel; (*D*) 5A Zeolite; and (*E*) activated carbon. The vapor pressure at 100% R.H. is 23.6 Torr (Yang, 1997).

the BDDT classification (Yang, 1997)) due to pore filling or capillary condensation in the micropores. Activated carbon is used, nonetheless, in processes dealing with humid gas mixtures and water solutions because the organic and nonpolar or weakly polar compounds adsorb more strongly, and hence preferentially, on its surface than water does.

A. Recent Developments on Activated Carbon

Interesting developments on activated carbon have been reported recently. They include chemical modification of the surfaces, activated carbon fibers (ACF), and CH_4 and H_2 storage. A brief discussion is given next.

Adsorption of water vapor on activated carbon has been studied extensively because of its scientific as well as practical importance. Chemical modification can significantly alter the adsorption behavior. It has long been known that oxidation and reduction affect the *hydrophobicity* of carbon. The water isotherm generally follows an S-shaped curve, with little or no adsorption at P/P_0 below 0.3 or 0.4. In this region, water molecules are bonded to certain oxygen complexes, likely by hydrogen bonding and electrostatic forces (the nonspecific interactions by Lennard–Jones 6–12 potential are insignificant). At higher P/P_0, clusters and eventually pore filling occur through hydrogen bonding. Pore structure comes into play only in the latter stage. Oxidation of the surface increases the oxygen complexes and hence shifts the threshold P/P_0 for water adsorption. The extensive literature on this subject has been discussed elsewhere (Jankowska, Swiatkowski, and Choma, 1991; Rouquerol, Rouquerol, and Sing, 1999; Leon y Leon and Radovic, 1992; Rodriquez-Reinoso, Molina-Sabio, and Munecas, 1992; Carrasco-Marin *et al.,* 1997; Salame and Bandosz, 1999).

Although the effects of oxidation on water adsorption are expected, the effects of incorporation with other atoms are quite unexpected. Mild chlorination of the surface of activated carbon results in slightly more hydrophobicity (Hall and Holmes, 1993). Mild fluorination has shown drastically increased hydrophobicity, even at $P/P_0 \approx 1$ (Kaneko, Ohbu *et al.,* 1995; Kaneko, Yang *et al.,* 2000). The work on fluorination by Kaneko and colleagues was done on active carbon fibers(ACFs), but their observations are expected on activated carbon as well. Incorporation of nitrogen atoms on ACFs, on the contrary, decreased the hydrophobicity (Kaneko, Yang *et al.,* 2000). A further understanding of the different effects by different chemical modifications is clearly needed.

Activated carbon fibers were a remarkable technological development. The ACFs have high mechanical strengths, high surface areas (≈ 1000 m^2/g), and microporosity (8-10 Å pore dimension) and can be formed into cloth.

High-strength carbon fibers have been produced since 1950s, but ACFs have been available commercially only recently. An activation step is necessary starting from the carbon fibers. Excellent reviews of the development of and studies on ACFs have been made by Suzuki (1994) and Rouquerol, Rouquerol and Sing (1999). Many possible novel applications of the ACF's have been reported (Suzuki, 1994; Kaneko, 2000).

Energy storage, i.e., storage of methane and hydrogen, has attracted much interest, particularly for onboard-vehicle applications. Activated carbon has been the most promising candidate as the sorbent for both methane and hydrogen storage.

Methane cannot be liquefied at ambient temperature ($T_c = 190.6$ K); hence, very high pressures (typically up to 25 MPa) are needed for the requirement of onboard storage. The pressure can be reduced by using sorbents. This subject has been investigated since the early 1980s (Talu, 1993). The target pressure for storage is 4 MPa. The interaction between methane and carbon is only by nonspecific dispersion forces. Among all activated carbons that have been investigated, the storage capacity is approximately proportional to the BET surface area. This proportionality is due to the wide pore-size distribution in the commercial activated carbons. Molecular simulations, however, showed that there is an optimal pore size (assuming slit-shaped pores) for maximum storage (Matranga, Myers, and Glandt, 1992; Jiang, Zollweg, and Gubbins, 1994). which is near 11 Å. Indeed, activated carbon fibers, which have relatively uniform pore sizes (near 10 Å), yielded higher methane-sorption capacities than activated carbons (Alcaniz-Monge *et al.,* 1997). The best carbons have capacities near 200 NTP v/v at 4 MPa at ambient temperature, which is quite adequate for onboard-vehicle applications. The effects of typical impurities that are contained in the natural gas on the charge–discharge behavior have been analyzed (Mota, 1999).

Interests in hydrogen storage by adsorption were intensified only recently. There are four possibilities for onboard hydrogen storage: compressed gas, liquefaction, metal hydrides, and adsorption. Limits on the first three are now well defined. This is not the case with adsorption. Recent claims on hydrogen adsorption in carbon nanotubes have stimulated intense interest (as well as suspicion), which will be discussed separately. The target of 5% (by weight) of storage capacity at 100 atm (and ambient temperature) has been set by the Department of Energy for sorbent development (Yang, 2000). With a typical commercial activated carbon, the amounts adsorbed at 293 K were 0.5% (weight) at 100 atm and 0.32% (weight) at 50 atm (Lamari, Aoufi, and Malbrunot, 2000). However, as environmental concerns and interest in fuel-cell technology increase, the target for hydrogen storage capacity will undoubtedly be lowered. With modifications, activated carbon or ACF could potentially meet the needs.

B. Activated Alumina and Silica Gel

Alumina, silica and many other metal oxides are insulators. However, recent experiments indicate that the surfaces of these insulators are mainly ionic (Masel, 1996). The pristine or freshly cleaved surfaces of single crystals of these oxides (cleaved under ultrahigh vacuum) are fairly inert and do not have significant adsorption capacities for even polar molecules such as CO and SO_2 (Masel, 1996; Henrich and Cox, 1994). However, the surface chemistry and adsorption properties are dominated by defects on real surfaces. For example, oxide vacancies on alumina expose the unsaturated aluminum atoms, which are electron acceptors, or Lewis acid sites.

The commercial alumina and silica gel sorbents are mesoporous, i.e., with pores mostly larger than 20 Å (see Fig. 1). Activated alumina is produced by thermal dehydration or activation of aluminum trihydroxide, Al $(OH)_3$ (Yang, 1997), and is crystalline. Commercially, silica is prepared by mixing a sodium silicate solution with a mineral acid such as sulfuric or hydrochloric acid. The reaction produces a concentrated dispersion of finely divided particles of hydrated SiO_2, known as silica hydrosol or silicic acid:

$$Na_2SiO_3 + 2HCL + nH_2O \rightarrow 2NaCl + SiO_2 \bullet nH_2O + H_2O. \tag{11}$$

The hydrosol, on standing, polymerizes into a white jellylike precipitate, which is silica gel. The resulting gel is washed, dried and activated. Various silica gels with a wide range of properties such as surface area, pore volume, and strength can be made by varying the silica concentration, temperature, pH, and activation temperature (Iler, 1979). Two typical types of silica gel are known as regular-density and low-density silica gels, although they have the same densities (true and bulk). The regular-density gel has a surface area of 750–850 m^2/g and an average pore diameter of 22–26 Å, whereas the respective values for the low-density gel are 300–350 m^2/g and 100–150 Å.

The silica gel is amorphous. Using high-resolution electron microscopy, it is known that its amorphous framework is made up of small globular (primary) particles having sizes of 10 to 20 Å (Rouquerol, Rouquerol and Sing, 1999). An alternative route involves reactions of silicon alkoxides with water, and a wide variety of materials can be made this way (Jones 1989; Brinker and Sherer, 1990). The processes based on this route are referred to as *sol-gel processing,* and they offer many promising possibilities. For silica gel, the reaction is

$$Si(OR)_4 + 2H_2O \xrightarrow{ROH} SiO_2 + 4ROH. \tag{12}$$

Silicic acids are also formed by hydrolysis:

$$-\overset{|}{\underset{|}{\mathrm{Si}}}-\mathrm{OR}+\mathrm{H_2O}\;=\;-\overset{|}{\underset{|}{\mathrm{Si}}}-\mathrm{OH}+\mathrm{ROH} \tag{13}$$

The silicic acids thus formed can then polymerize via

$$-\overset{|}{\underset{|}{\mathrm{Si}}}-\mathrm{OH}+\mathrm{HO}-\overset{|}{\underset{|}{\mathrm{Si}}}-=-\overset{|}{\underset{|}{\mathrm{Si}}}-\mathrm{O}-\overset{|}{\underset{|}{\mathrm{Si}}}-+\mathrm{H_2O} \tag{14}$$

or

$$-\overset{|}{\underset{|}{\mathrm{Si}}}-\mathrm{OH}+\mathrm{RO}-\overset{|}{\underset{|}{\mathrm{Si}}}-=-\overset{|}{\underset{|}{\mathrm{Si}}}-\mathrm{O}-\overset{|}{\underset{|}{\mathrm{Si}}}-+\mathrm{ROH} \tag{15}$$

The reaction products are high-molecular-weight polysilicates (a sol), which form a three-dimensional porous network filled with solvent molecules (a gel). The more recent development of MCM-41 (to be discussed separately) is a derivative of the sol-gel route. The pore structure, as well as the surface chemistry, can be tailored in the sol-gel route. The pH value in the initial stages (Reactions 13–15) is a main factor in controlling the pore dimensions. Low pH (e.g., by adding HCl) leads to microporosity, whereas high pH (e.g., by adding ammonium hydroxide) results in mesoporosity (Brinker and Sherer, 1990). Apparently, pH influences the size distribution of the globular, primary particles and also how these particles agglomerate and, hence, the final pore structure. Sol-gel processing is highly versatile. It is not limited to silica; it is also applicable to many other main-group and transition metal oxides.

IV. MCM-41

Beck *et al.* (1992) succeeded in the synthesis of a new family of ordered, mesoporous silicate/aluminosilicate by hydrothermal formation of silica gels in the presence of surfactant templates. Typically, quaternary ammonium surfactants were used. The surfactants self-assemble to form micellar templates with a three-dimensional, long-range order. The silicate precursors condense on the walls of the template. The organic templates are subsequently removed by air oxidation, leaving behind a silicate structure. These materials are amorphous but exhibit simple X-ray diffraction patterns ($2\theta = 2^\circ$) that reflect the interplanar spacing of the regular mesoporous structure of the templates. These materials were named M41S, and the materials that have the honeycomb-shaped structures are named MCM-41. A schematic of the formation of MCM-41 is shown in Fig. 3.

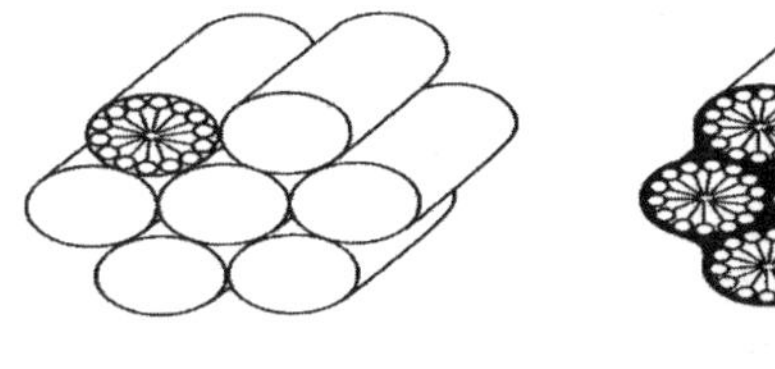

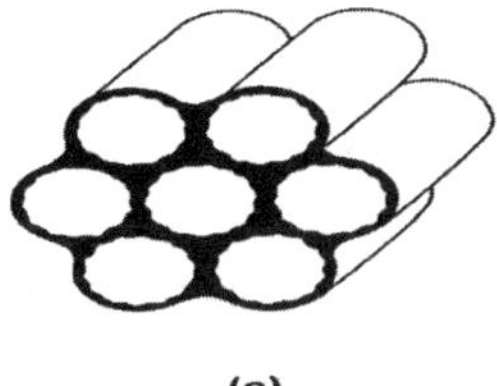

FIG. 3. Schematic representation of the formation of MCM-41 by the liquid-crystal templating mechanism. (a) Hexagonal array of cylindrical micelles; (b) the same, with silicate species between the cylinders; and (c) hollow cylinders of MCM-41 after thermal elimination of organic material. (Rouquerol, Rouquerol and Sing, 1999).

Different mechanisms for the interactions between the silicate precursors and the organic template as well as the liquid crystal templating mechanisms have been discussed (Ying, Mehnert, and Wong, 1999; Tanev and Pinnavaia, 1995). Two other major types of M41S materials are MCM-48 (with 3-D pores) and MCM-50 (with a pillared layer structure).

Among the M41S materials, MCM-41 has received most attention because of its simple structure as well as the ease of synthesis and tailoring of its structure and surface properties. The pore dimension is in the range of 20 to 100 Å and can be tailored by several different strategies (Rouquerol, Rouquerol and Sing, 1999; Ying, Mehnert, and Wong, 1999; Zhao, Lu, and Millar, 1996). The first one is to use surfactants with different chain lengths, and pore diameters can be controlled from near 15 A to 45 Å (Huo, Mragolese, and Stucky, 1996). The use of two surfactants can extend the pore sizes to 55 Å (Kaman, Anderson, and Brinker, 1996). The addition of expander molecules such as trimethylbenzene (Beck *et al.*, 1992; Zhao, Lu, and Millar, 1996) extends the pore size to near 100 Å. Good quality, large pore MCM-41 with pore sizes up to 65 Å can be made by postsynthesis hydrothermal restructuring (Huo, Mragolese, and Stucky, 1996; Sayari *et al.*, 1997). Ordered, nanostructured materials with even larger pores (up to 300 Å) can be obtained by using block copolymers as the structure-directing agents (Zhao *et al.*, 1998). Moreover, MCM-41 with heteroatoms (such as Ti, B and V) and many non-silica materials with structures similar to MCM-41 can be synthesized (Ying, Mehnert, and Wong, 1999; Sun and Ying, 1997).

The unique feature of MCM-41 is the large uniform pore structure (and hence remarkably large pore volume, above 0.7 cm^3/g). This unique feature has made these materials promising for catalysis (as discussed by Ying and colleagues (1999)). However, this unique feature is not attractive for gas adsorption because the interaction potentials are not enhanced within the pores, as discussed earlier. Consequently, relatively few studies have been made on MCM-41 as adsorbents.

In order to obtain useful adsorption properties, MCM-41 needs to be modified in either surface chemistry or pore structure. The reported modifications have been based mainly on the use of reactive silanes that contain organic groups (such as alkyls) and chloride. The formed MCM-41 has hydroxyl groups, and by reacting the hydroxyl group with silanes, i.e. silanation, pore sizes can be reduced by the grafted silanes (Feng *et al.,* 1997; Jaroniec *et al.,* 1998). The grafting can be accomplished by many other compounds such as metal alkoxides and halides, hence the chemistry of the surface can be altered or functionalized (Moller and Bein, 1998; Jaroniec *et al.,* 1998). Functionalization of the surface can be also accomplished before the final calcination step by directly displacing the sulfactants with reactive silanes (Antochsuk and Jaroniec, 2000). A technique to reduce the pore size only at the opening regions has been suggested (Zhao, Lue, and Hu, 1999), by grafting a controlled number of layers of silica in the opening regions. This was done by displacing the surfactant in the opening regions by H^+, followed by silanation with $Si(OEt)_4$ and hydrolysis, and finally calcination.

Only few promising applications of the MCM-41 and its modified forms have been reported. Izumi and coworkers have reported the use of MCM-41 for VOC removal (Izumi, 1996) and SO_2 removal (Teraoka *et al.,* 2000), by taking advantage of the weak bonds with the surface and, hence, ease in desorption. They also reported a low-temperature synthesis route for MCM-41 at very low pH (<1), which is a possible low-cost method. Feng *et al.* (1997) grafted the MCM-41 with a silane-containing thiol (—SH) group and produced a sorbent that is highly selective for binding heavy metal ions such as mercury, silver, and lead ions from wastewaters. The sorbent is also regenerable with HCl. Because of the large arrays of functionalities and pore sizes that can be achieved with the MCM-41 material, unique adsorption properties will no doubt be obtained. However, due to the cost, its use appears to be limited to specialized applications.

V. Zeolites

Zeolites are crystalline aluminosilicates of alkali or alkali earth elements such as sodium, potassium, and calcium and are represented by the stoichiometry:

$$M_{x/n}\,[(AlO_2)_x(SiO_2)_y]\,zH_2O,$$

where x and y are integers, with $y/x \geq 1$, n is the valence of cation M, and z is the number of water molecules in each *unit cell.* Unit cells are shown in Fig. 4(b) and (c). The cations are necessary to balance the electrical charge

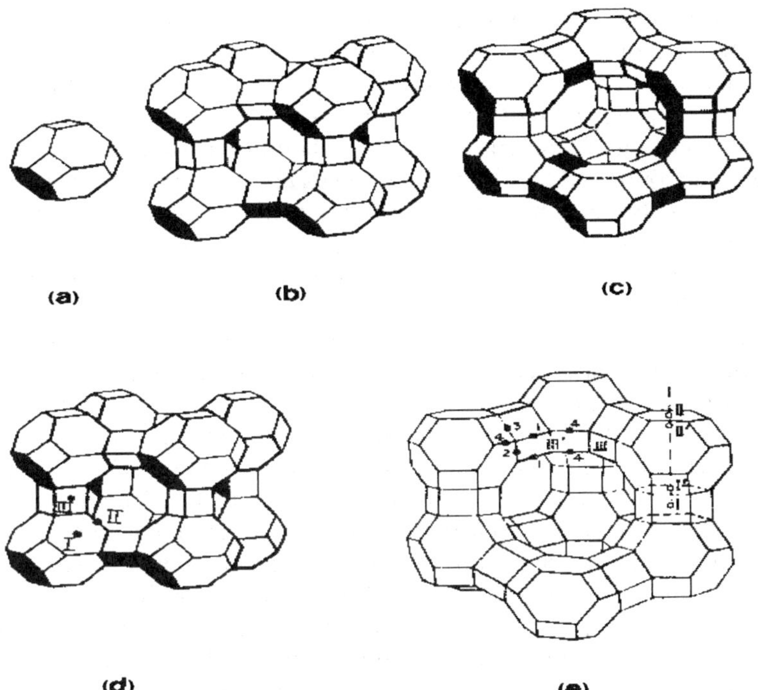

FIG. 4. Line representations of zeolite structure: (a) sodalite cage, or truncated octahedron; (b) type A zeolite unit cell; (c) unit cell of types X and Y, or faujasite; (d) cation sites in type A (there are 8 I, 3 II, and 12 III sites per unit cell); (e) cation sites in types X and Y (16 I, 32 I′, 32 II, 32 II′, 48 III, and 32 III′, sites per unit cell).

of the aluminum atoms, each having a net charge of –1. The water molecules can be removed with ease upon heat and evacuation, leaving an almost unaltered aluminosilicate skeleton with a void fraction between 0.2 and 0.5. The skeleton has a regular structure of cages, which are usually interconnected by six windows in each cage. The cages can imbide or occlude large amounts of guest molecules in place of water. The size of the window apertures, which can be controlled by fixing the type and number of cations, ranges from 3 Å to 10 Å . The sorption may occur with great selectivity because of the size of the aperature (and to a lesser extent because of the surface property in the cages)—hence the name *molecular sieve.*

At least 40 species of naturally occurring zeolites have been found. The principal ones are chabazite, $(Ca, Na_2)Al_2Si_4O_{12}(6\,H_2O)$; gmelinite, $(Na_2, Ca)Al_2Si_4O_{12}(6\,H_2O)$; mordenite, $(Ca, K_2, Na_2)Al_2Si_{10}O_{24}(6.66\,H_2O)$; levynite, $Ca,Al_2Si_3O_{10}(5\,H_2O)$; and faujasite, $(Na_2, Ca, Mg, K_2)OAl_2Si_{4.5}O_{12}\,(7\,H_2O)$. More than 150 types of zeolites have been sythesized; they are designated by a letter or group of letters: Type A, Type X, Type Y, Type ZSM, and so on. The commercial production of synthetic zeolites started with the successful development of low-temperature (25–100°C) synthesis methods using very reactive materials such as freshly coprecipitated gels or amorphous solids (Breck, 1974; Milton, 1959). Two comprehensive monographs, by Barrer (1978) and Breck (1974), deal with all aspects of zeolites. The zeolites that have been synthesized more recently are discussed by Szostak (1998).

A. Structures and Cation Sites

The primary structural units of zeolites are the tetrahedra of silicon and aluminum, SiO_4 and $A1O_4$. These units are assembled into secondary polyhedral building units such as cubes, hexagonal prisms, octahedra, and truncated octahedra. The silicon and aluminum atoms, located at the corners of the polyhedra, are joined by a shared oxygen. The final zeolite structure consists of assemblages of the secondary units in a regular three-dimensional crystalline framework. The ratio Si/Al is commonly 1 : 5. The aluminum atom can be removed and replaced by silicon in some zeolites, thereby reducing the number of cations, and the cations can also be exchanged. The inner atoms in the windows are oxygen. The size of the windows depends, then, on the number of oxygen atoms in the ring—four, five, six, eight, ten or twelve. The aperture size, as well as the adsorptive properties, can be further modified by the number and type of exchanged cations. A description of the structures will be given only for the zeolites important in gas separation—Type A and Types X and Y.

1. Type A

The structural unit in Type A zeolite, as well as that in Types X and Y, is the truncated octahedron, shown in Fig. 4(a). This unit is also called a sodalite cage, because sodalite is formed by directly fusing the four-member rings of the units. The four-member rings of the sodalite units can also be linked through four-member prisms, as shown in Fig. 4(b), which is Type A zeolite. The unit cell of Type A zeolite, as shown in this figure, contains 24 tetrahedra, 12 AlO_4, and 12 SiO_4. When fully hydrated, 27 water molecules are contained in the central cage, or cavity, of the unit cell and in the eight smaller sodalite cages. The free diameter in the central cavity is 11.4 Å, which is entered through six eight-member oxygen-ring apertures with a minimum diameter of 4.4 Å. There are 12 negative charges to be balanced by cations in each unit cell. The most probable locations for the cations are indicated in Fig. 4(d). Type I is at the center of the six-member ring and thus at one of the eight corners of the cavity. Type II is at the eight-member aperture, directly obstructing the entrance. Type III is near the four-member ring inside the cavity.

Type A zeolites are synthesized in the sodium form, with 12 sodium cations occupying all eight sites in I and three sites in II, plus one site in III. This is the commercial Type 4A zeolite, with an effective aperture size of 3.8 Å. The sodium form can be replaced by various other cations or by a hydrogen ion. The commercial Type 3A zeolite is formed by exchanging Na^+ with K^+, resulting in a smaller effective aperture size due to the larger K^+. The aperture size of the sodium form can also be increased by exchanging Na^+ with Ca^{+2} or Mg^{+2}, because 2 Na^+ are replaced by one divalent cation. The form of the exchanged Ca^{+2} or Mg^{+2} is Type 5A with rather unobstructed and larger apertures.

2. *Types X and Y*

The skeletal structure of Types X and Y zeolites is the same as that of the naturally occurring faujasite. The sodalite units are linked through six-member prisms, as shown in the unit cell in Fig. 4(c). Each unit cell contains 192 $(Si, Al)O_4$ tetrahedra. The number of aluminum ions per unit cell varies from 96 to 77 for Type X zeolite and from 76 to 48 for Type Y zeolite. This framework has the largest central cavity volume of any known zeolite, amounting to about 50% void fraction in the dehydrated form. A unit cell, when fully hydrated, contains approximately 235 water molecules, mostly in the central cavity. The aperture is formed by the 12-member oxygen rings with a free diameter of approximately 7.4 Å. Three major locations for the cations are indicated in Fig. 4(e). The locations are center of the six-member prism (I) and opposite to I in the sodalite cage (I′); similar to I and I′ but further from the central cavity (II and II′); and the 12-member aperture (III and III′). The commercial 10X zeolite contains Ca^{+2} as the major cation, and Na^+ is the major cation for 13X zeolite. The distribution of Na^+, K^+, Ca^{+2}, other cations, and H_2O among the sites in X and Y zeolites has been discussed in detail by Barrer (1978). The BET surface are a measured with N_2 for zeolites falls in the range between 500 and 800 m^2/g.

B. Unique Adsorption Properties: Anionic Oxygens and Isolated Cations

Zeolites exhibit many unique adsorption properties, mainly because of their unique surface chemistry. The surface of the framework is essentially oxygen atoms, because Si and Al are buried or recessed in the tetrahedra of oxygen atoms, so they are not exposed and cannot be accessed by adsorbate molecules. Also, the anionic oxygen atoms are much more polarizable than the Al and Si cations, as well as being more abundant. Therefore, the numerous and anionic oxygen atoms dominate the van der Waals interactions with the sorbate molecules, i.e., $\phi_D + \phi_R$(dispersion + repulsion).

Besides the anionic oxygen, cations are located at certain sites, and most of these sites are hidden or inaccessible to the adsorbate molecules. However, with molecules with permanent dipoles and quadrupoles, the interactions with these few exposed cations can dominate the total interaction potential.

The anionic surface oxygens carry negative charges, and the charge depends on the location of the oxygen relative to the cation sites and also on the cation. In Monte Carlo simulations, a constant charge is usually assigned to all surface oxygen atoms and the value is usually determined by fitting the experimental data (of isosteric heat of adsorption or the isotherm). For example, a charge of −1/3 was used by Razmus and Hall (1991) and −1.2 was used by Mellot and Lignieres (1997).

It is instructive to compare the relative electronegativities of the anion of the zeolite framework with simple anions such as halides. Such a comparison can be made by calculating the net charges of the anions (or cations) using molecular orbital theories. The Gaussian 94 Program in Cerius^2 molecular modeling software from Molecular Simulations, Inc., was used for the calculation (Takahashi, Yang, and Yang, 2000; Yang and Yang, unpublished). The calculations were performed at the Hartree–Fock (HF) and density functional theory (DFT) levels using effective core potentials (ECPs). The LanL2DZ basis set was employed for both geometry optimization and natural bond orbital (NBO) analysis. The net charges were calculated by using NBO. The zeolite model used in the calculation was the cluster model shown in Fig. 5. The results are shown in Table V.

From Table V, the zeolite anion is more electronegative than F^-. Also shown in Table V are the electron occupancies in the 5*s* orbital of the Ag^+ that is bonded to the anion. For a perfect anion, the Ag^+ to which it is bonded should have an empty 5*s* orbital. Again, it is seen that Ag^+ in AgZ has the lowest occupancy in its 5*s* orbital, indicating that the Z^- (i.e., zeolite framework anion) is the most electronegative anion.

The strong anionic nature of the zeolite framework and the corresponding, strong cations are unique with zeolites. Furthermore, the cations and anions are not located closely to each other, which exert strong electric fields over the surface.

C. Interactions with Cations: Effects of Site, Charge, and Ionic Radius

In earlier discussions, the strong or dominating contributions of cation–dipole and cation–quadrupole interactions to the total bonding were seen

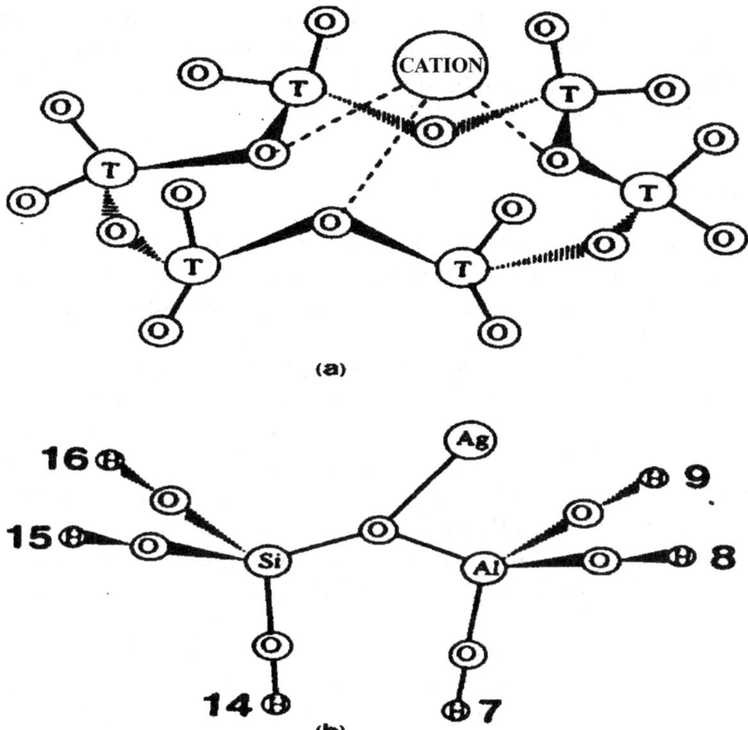

FIG. 5. (a) Site II cation on a six-membered oxygen ring as the basic unit on types A and X zeolites. T denotes Si or Al. (b) Geometry-optimized cluster model to represent the chemistry of Ag/zeolite.

TABLE V
RELATIVE ELECTRONEGATIVITIES OF ZEOLITE ANION AND HALIDES—COMPARISON OF ANION NET CHARGES CALCULATED BY MOLECULAR ORBITAL THEORY USING B3LYP/LANL2DZ WITH NBO METHOD[a]

	Anion charge, electronic unit	Electron occupancy in 5*s* orbital of Ag^+
Ag^+Z^-	0.5765	0.142
Ag^+F^-	0.5111	0.295
Ag^+Cl^-	0.3404	0.357
Ag^+Br^-	0.3017	0.393
Ag^+I^-	0.2375	0.437

[a]AgZ denotes Ag zeolite using the model in Fig. 5.

(see Table I). Unfortunately, except ionic radii, the cation sites and the charges have not been well determined. Hence they remain largely as fitting parameters in molecular simulations. However, the strong effects of these parameters on adsorption are well established.

1. Effects of Cation Sites

Cation sites are well defined in type A zeolite but are not well defined in nearly all other zeolites, including LSX (low silica X, Si/Al $=1$). Most of the cation sites are hidden and are not accessible for interactions with the adsorbate molecules.

Effects of cation sites can be best illustrated with the important system of N_2/O_2 on type X zeolites. Na/X (or 13X) has been used commercially for air separation since the 1970's. Li/LSX is the best sorbent commercially available today (Chao, 1989). Mixed-cation AgLi/LSX (with 1–3% Ag cations) has been shown to be even better than Li/LSX (Yang and Hutson, 1998; Hutson, Rege, and Yang, 1999).

As shown in Fig. 4, there are 192 possible cation sites in a unit cell of faujasite (or X zeolite). For LSX, there are only 96 cations (monovalent). Upon activation of the zeolite, i.e., heating at 350°C, the cations migrate to the sites with the lowest energies. Migration is an activated process, which depends on the temperature, time, and the size of the cation. Unfortunately, the most stable sites (that have the lowest energies) are the hidden sites, not exposed to the supercage cavity. So relatively few exposed sites are occupied.

By ion exchange of Na^+ with Li^+ in the LSX, Chao (1989) obtained significantly improved N_2/O_2 selectively. This improvement was the result of the smaller ionic radius of Li^+ (0.68 Å) compared to Na^+ (0.97 Å), with equal charge and, hence, much higher $\phi_{\dot{F}Q}$ (electric field gradient–quadrupole) potential. However, no improvement was seen until approximately more than a 70% ion exchange was made, and the selectivity increased linearly with ion exchange beyond this threshold value (see Fig. 6). (Figure 6 actually shows LiX with different Si/Al ratios, or different number of Li cations per unit cell. But it illustrates the same phenomenon.) The reason for this significant phenomenon was that Sites I, I′ and II′ were lower-energy sites, which were preferred by Li^+ (Chao, *et al.*, 1992; Coe, 1995). Sites II and III are exposed sites but have lower coordination and are less preferred. These exposed sites are most important for adsorption.

Although Ag^+ has a larger ionic radius (1.26 Å) than Li^+, we have found that π-complexation bond was formed between Ag^+ (in AgZ) and N_2 (Chen and Yang, 1996). This π-complexation bond, although weak, enhanced the adsorption for N_2 significantly (Yang and Hutson, 1998; Hutson, Rege, and Yang, 1999).

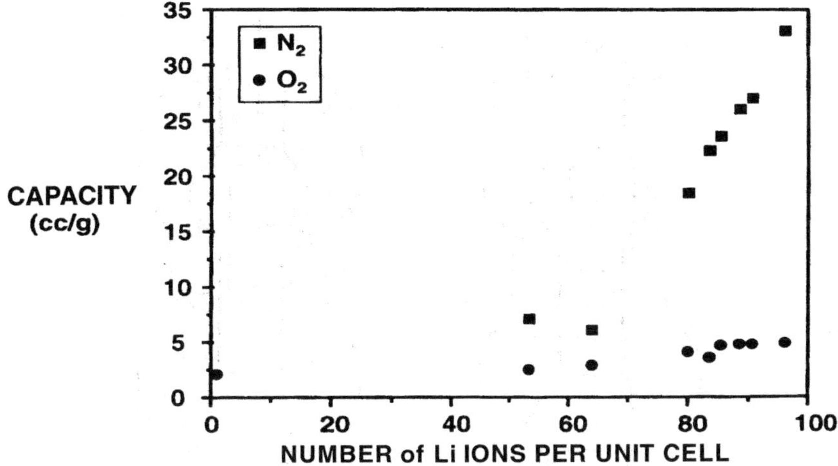

FIG. 6. N_2 and O_2 adsorption capacities at 23°C and 1 atm for Li faujasite with different Si/Al ratios. (Coe, 1995; this result is similar to that given in Chao (1989)).

The pure Ag/LSX (Si/Al = 1) adsorbs 22 nitrogen molecules per unit cell at 1 atm and 25°C. The capacity depends on the temperature of heat treatment, as shown in Fig. 7. X-ray photoelectron spectroscopy (XPS) results showed that some reduction takes place during heating from 350 to 450°C in *vacuo* or in an inert atmosphere. Moreoever, a color change

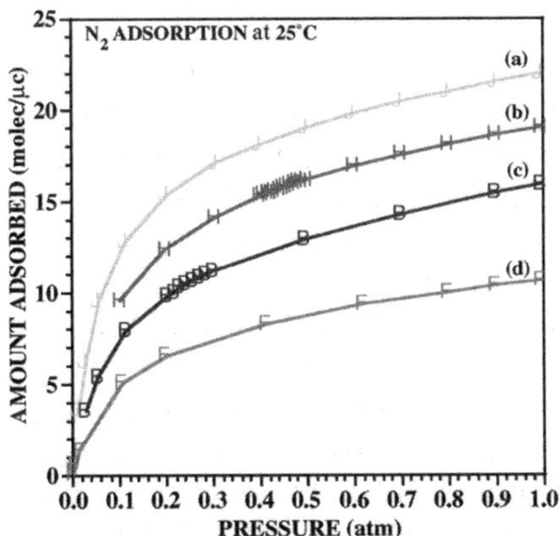

FIG. 7. N_2 adsorption isotherms, measured at 25°C, for Ag/LSX (a) after drying at room temperature followed by vacuum dehydration at 450°C, (b) after drying at room temperature followed by vacuum dehydration at 350°C, (c) after drying in air at 100°C followed by vacuum dehydration at 350°C and (d) after drying in air at 100°C in air, followed by heat treatment in air at 450°C and, finally, vacuum dehydration at 450°C (Hutson and Yang, 2000).

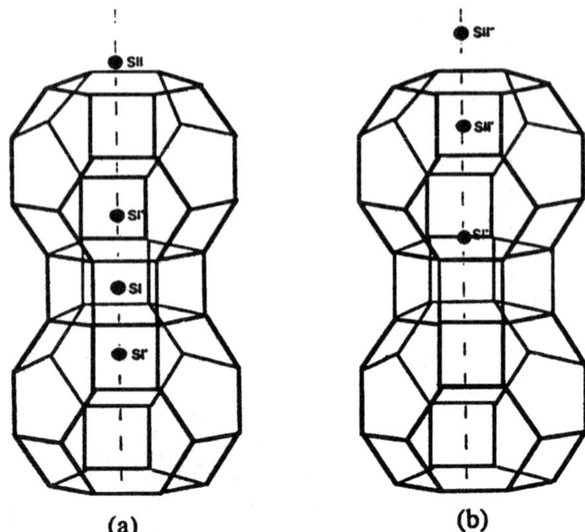

FIG. 8. Sodalite and hexagonal prism site I and II cation locations in Ag/faujasites. Configuration (a) shows the normal cation locations with occupied sites at SI, SI′ and SII. Configuration (b) shows cation sites that have resulted from cation and/or cluster migration upon vacuum thermal treatment. This configuration shows occupied sites at SI′*, SII′, and SII* (Hutson and Yang, 2000).

from white to brick red occurs. These are the result of the formation of trinuclear Ag^+–Ag^0–Ag^+ cluster. A detailed neutron diffraction analysis has identified the site of the Ag cluster as shown in Fig. 8. From Fig. 8, it is seen that some of the Ag^+ originally located at site SII (after heating to 350°C) are now located at site SII*. The cation at SII is significantly shielded by the six oxygen atoms o the 5-ring and hence sterically is only partially accessible to the adsorbate N_2. After heating to 450°C, the Ag^+ located at SII* becomes less shielded by the 6 O atoms. Hence, there are more interactions (including weak π-complexation) with nitrogen.

The isosteric heats of adsorption of N_2 on Li/LSX and Ag/LSX are shown in Fig. 9.

In is seen that the first N_2 molecule adsorbed in the unit cell of AG/LSX has a bond energy of about 10 kcal/mol, decreased quickly to below 7 kcal/mol for other N_2 molecules. This difference of 3 kcal/mol is the result of bonding with Ag^+ at Site II*. The vertical distance between SII and SII* is 0.75 Å (Fig. 8). This small difference causes significantly less shielding (by O atoms) and consequently much stronger bonding with the sorbate molecule.

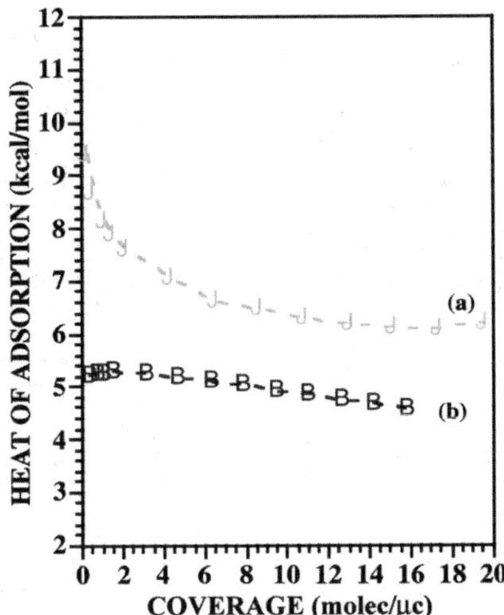

FIG. 9. Heat of adsorption (kcal/mol) versus surface coverage (molec/μc) for N_2 adsorption on (a) Ag-LSX-450 and (b) Li LSX-450.

Another example for illustrating the cation-shielding effect on adsorption is by comparing the N_2 adsorption on NaY and LiY zeolites (Mellot and Lignieres, 1997). The type Y zeolite has the same framework structure as X but with less than 76 cations per unit cell (due to higher Si/A1 ratios). In this case, Site II is the only exposed site for Na^+ and Li^+. LiY is expected to adsorb N_2 more strongly than NaY because of its smaller cations. The experimental isotherms are, however, the same (Mellot and Lignieres, 1997). The reason is that the Li^+ in site II is more shielded by O atoms, as evidenced by a shorter Li^+–O (framework) bond (of 2.07 Å) than the Na^+–O (framework) bond (of 2.48 Å) (Shepelev, Anderson, and Smolin, 1990). Moreover, on the same X zeolite framework, it was necessary to assign different charges for the four Ca^{2+} cations (2, 1.2, 1.2, 1.2) on the four SII sites in order to account for the energetic heterogeneity for N_2 adsorption (Mellot and Lignieres, 1997).

2. *Effects of Cation Charge and Ionic Radius*

The equilibrium distance between an interacting pair is the sum of the van der Waals or ionic radii of two atoms. Hence the ionic radius of the cation is important in all interactions, both nonspecific and electrostatic interactions. The ionic radii of important cations are listed in Table III. The cationic charge, on the other hand, is important only to the electrostatic interactions.

The effects of cation charge and ionic radius on the interaction energies are best seen in Eqs. (4)–(8). For electrostatic interactions, the following dependence holds:

Induction:

$$\phi_{\text{Ind}} \propto \frac{q^2\alpha}{r^4}. \tag{16}$$

Field-dipole:

$$\phi_{F\mu} \propto \frac{q\mu}{r^2}. \tag{17}$$

Field gradient-quadrupole:

$$\phi_{\dot{F}Q} \propto \frac{qQ}{r^3}, \tag{18}$$

where

$$r = r_i \text{ (ionic radius)} + r_j \text{(adsorbate atom or molecule)}. \tag{19}$$

Table VI shows the interaction energies of Ar, O_2 and N_2 interacting with isolated cations. For Ar, Eqs. (4)–(8) were used (Barrer and Stuart, 1959). For O_2 and N_2, the values were calculated from quantum mechanics which represent the sum of L–J and electrostatic interactions (Mellot and Lignieres,

TABLE VI

INTERACTION ENERGIES (ϕ) BETWEEN MOLECULES AND ISOLATED CATIONS[a]

Molecule or Ion	r Å	$10^{24}\alpha$ cm^3	Q esu	$-(\phi_D + \phi_R)$ kJ/mol	$-\phi_{\text{Ind}}$ kJ/mol	$-\phi_{\text{total}}$ kJ/mol
Ar	1.92	1.63	0			
O(O_2)	1.73	1.58	−1.3			
N(N_2)	1.89	1.74	−4.7			
Ar-Ion						
Li^+	0.78	0.029	0	0.210	21.3	21.5
Na^+	0.98	0.180	0	0.670	16.0	16.6
K^+	1.33	0.840	0	1.80	10.2	12.0
Ca^{2+}	0.99	0.471	0	2.13	63.5	65.6
Sr^{2+}	1.13	0.863	0	3.26	52.7	55.9
Ba^{2+}	1.35	1.560	0	4.22	40.7	44.9
O_2-Li^+						32
O_2-Na^+						20
N_2-Li^+						51
N_2-Na^+						36

[a]Values for O_2 and N_2 are from Mellot and Lignieres (1997) and all others are from Barrer (1978); Van der Waals radius and ionic radius are denoted by r. N_2-ion and O_2-ion are in linear arrangements.

TABLE VII
COMPONENTS OF INTERACTION ENERGIES (ϕ, IN kJ/mol) FOR CO_2 ADSORBED ON X ZEOLITE WITH DIFFERENT CATIONS[a]

Component	Li	Na	K	Rb	Cs
$-\phi_D$ (Oxygens)	15.9	13.0	7.1	4.6	4.6
$-\phi_D$ (Cations)	0.4	0.8	3.3	4.6	9.2
$-\phi_{\text{Ind}}$	9.6	5.0	2.1	0.8	0
$-\phi_{\dot{F}Q}$	30.9	21.3	17.6	14.6	9.6

[a] CO_2 is oriented along the ppp axis in the cavity (Barrer, 1978; Barrer and Gibbons, 1965); for CO_2: $\mu = 0$ and $Q = -4.3$ ESU.

1997). For Ar-cation pairs, as the cation increases in size, the polarizability increases, and so ($\phi_D + \phi_R$) also increases. The induction energy, in contrast, decreases sharply with the increasing size, as shown in Eq. (16) ($\propto r^{-4}$). The divalent cations are slightly bigger but have twice the amount of charge; hence, the induction energy increases sharply ($\propto q^2$). For N_2 and O_2 interacting with the same cation, the nonspecific ($\phi_D + \phi_R$) and ϕ_{Ind} energies are about the same because their sizes, polarizabilities, and magnetic susceptibilities are about the same. The main difference in the total interacting energies comes from $\phi_{\dot{F}Q}$. The substantial differences for the four pairs (O_2–Li^+, O_2–Na^+, N_2–Li^+ and $N_2 - Na^+$) are clearly seen in Table VI. For O_2 interacting with Li^+ and Na^+, the difference of 12 kJ/mol is caused by the different sizes of the ions (see Eq. (18)). For the same ion with O_2 and N_2, the large difference is caused by the difference in the quadrupole moment. The dependence follows Eq. (18).

Barrer and Gibbons (1965) did calculations for the interaction potentials of CO_2 and NH_3 moving along the axes running through the center of the 12-ring window of faujasite-type 26-hedral cage of zeolite X. The results are shown in Tables VII and VIII. The qualitative comparison with experimental

TABLE VIII
ENERGY TERMS IN kJ/mol FOR NH_3 IN X ZEOLITE WITH DIFFERENT CATIONS (BARRER, 1978; BARRER AND GIBBONS, 1965); FOR NH_3: $\mu = 1.47$ DEBYE AND $Q = -1.0$ ESU

Cation	$-\phi_D$ (Oxygens)	$-\phi_D$ (Cations)	$-\phi_R$	$-\phi_{F\mu}$	$-\phi_{\text{Ind}}$	$-\phi_{\text{total}}$	Expt'l ($-\Delta$H)
Li	47.2	1.7	41.3	50.6	23.0	77.7	76.5
Na	33.8	3.3	23.4	33.0	9.6	53.9	72.3
K	9.6	6.7	14.6	20.1	3.8	23.8	59.8
Rb	7.9	11.7	18.0	17.6	2.9	20.5	55.6
Cs	7.5	16.3	19.6	15.0	2.1	19.6	47.2

data was remarkably good considering the calculations were made about 1965. In Table VII, the quadrupole-field gradient interaction dominates the adsorption of CO_2, because CO_2 has no dipole but has a strong quadrupole. The interaction energy is nearly proportional to r^{-3} (Eq. (18)), showing the strong dependence on the ionic radius of the cation. For NH_3, which has strong dipole and moderately strong quadrupole, the dipole-field interaction is clearly not important (Table VIII). The $\Phi_{F\mu}$ term is proportional to r^{-2} (Eq. (17)). The strong dependence on the cation size is also clearly seen in Table VIII.

A note needs to be made on the interactions with the zeolites with divalent (and higher-valent) cations. The energies of CO_2 on X zeolites with different univalent cations follow the order that larger ions give lower initial heats of adsorption (Table VII and (Barrer, 1978)). For divalent ions, the heats follow the reverse sequence of $Ba^{2+} > Sr^{2+} > Ca^{2+}$ (Barrer, 1978). This is also the case with N_2 adsorption (Mckee, 1964). Both N_2 and CO_2 are nonpolar but highly quadrupolar. For zeolites exchanged with univalent cations, the $\phi_{\dot{F}Q}$ term dominates. For divalent cations, however, the large polarizabilities (Table II) become important, and the dispersion and induction energies are large, especially for Ba^{2+}. Hence all interaction terms need to be considered.

VI. π-Complexation Sorbents

All industrial adsorption processes are essentially based on van der Waals and electrostatic interactions between the sorbate and the sorbent. Chemical bonds have yet to be exploited. As suggested by King (1987), chemical complexation bonds are generally stronger than van der Waals interactions (thus giving rise to higher selectivities), yet they are weak enough to be reversible (i.e., to be broken by simple engineering means). This picture is well illustrated by the bond-energy-bond-type diagram of G. E. Keller and colleagues (1992). Indeed, a number of important separations have been proposed by King and coworkers using solvents with functional groups to form reversible chemical complexation bonds between the solute and solvent molecules (King, 1987). The π-complexation is a special class complexation. It pertains to the main group (or *d*-block) transition metals. These metals and their ions can form the usual σ bonds with their *s* orbitals and, in addition, their *d* orbitals can back-donate electron density to the antibonding π orbitals of the molecule that is to be bonded. The π-complexation has been seriously considered for olefin/paraffin separation and purification by employing liquid solutions containing silver (Ag^+) or cuprous (Cu^{+1}) ions

(Quinn, 1971; Ho *et al.*, 1988; Keller *et al.*, 1992; Blytas, 1992; Eldridge, 1993; Safarik and Eldridge, 1998). Although gas–solid operations can be simpler as well as more efficient, particularly by pressure-swing adsorption, the list of attempts for developing solid π-complexation sorbents is a short one (Hirai; Hara, and Komiyama, 1985; Hirai, Kurima, *et al.*, 1985). CuCl, which is insoluble in water, has been considered for olefin–paraffin separations (Gilliand, Bliss, and Kip, 1941; Long, 1972). The only apparently successful solid sorbent of this nature before our work was $CuCl/\gamma\text{-}Al_2O_3$ for binding with the π bond of CO (Xie and Tang, 1990; Kumar *et al.*, 1993). It should also be noted that the commercially available sorbents do not have significant selectivities for olefins (over corresponding paraffins) and the use of these sorbents would require additional, substantial operations (Kulvaranon, Findley, and Liapis, 1990; Jarvelin and Fair, 1993; Ghosh, Lin, and Hines, 1993).

Efficient solid π-complexation sorbents have been developed within the last 5 years in the author's laboratory (Yang and Kikkinides, 1995; Chen and Yang, 1996; Wu *et al.*, 1999; Rege, Padin, and Yang, 1998; Huang, Padin, and Yang, 1999a, 1999b; Padin, Yang, and Munson, 1999; Padin and Yang, 2000), mainly for olefin–paraffin separations. The bond between the sorbent and the sorbate needs to be strong. However, excessively strong bonds lead to either reaction or irreversible adsorption. Empirically, the adsorption is "reversible" when the bond is below 15 to 20 kcal/mol, i.e., desorption can be achieved easily by simple engineering operations such as pressure and temperature changes. The strength of the bonding between sorbate and sorbent depends on

- Emptiness of the outer-shell *s* orbital of the cation on the sorbent surface
- The amount of π electrons in the target adsorbate molecule and the ease with which to donate to the cation

Molecular orbital theory has been used to study π-complexation (Chen and Yang, 1996; Huang, Padin, and Yang, 1999a, 1999b). Molecular orbital theory can serve as an ideal tool for designing sorbents for π-complexation for a given target adsorbate molecule.

A. π-Complexation Sorbents for Olefin–Paraffin Separations

Olefin–paraffin separations represent a class of most important and also most costly separations in the petrochemical industry. Cryogenic distillation has been used for more than 60 years for these separations (Keller *et al.*, 1992). They remain to be the most energy-intensive distillations because of the close relative volatilities. The most important olefin–paraffin separations are for the binary mixture of ethane–ethylene and propane–propylene. A number of alternatives have been investigated (Eldridge, 1993; Safarik and Eldridge, 1998). The most promising alternative is separation based on π-complexation.

In our laboratory, several new sorbents based on π-complexation were prepared for selective olefin adsorption. These include Ag^+-exchanged resins (Yang and Kikkinides, 1995; Wu *et al.*, 1999), monolayer $CuCl/\gamma\text{-}Al_2O_3$ (Yang and Kikkinides, 1995), monolayer CuCl on pillared clays (Cheng and Yang, 1995), and monolayer $AgNO_3/SiO_2$ (Rege, Padin, and Yang, 1998; Padin and Yang, 2000), and monolayer $AgNO_3$ supported on other substrates (Padin and Yang, 2000). The π-complexation bonds are reasonably understood through molecular orbital studies (Huang, Padin, and Yang, 1999a, 1999b). Among the different sorbents, monolayer $AgNO_3$ appears to be the best. Monolayer spreading of salts on high-surface-area substrates can be accomplished in several ways. One of the methods selected for study is known as *spontaneous monolayer dispersion.* Thermal monolayer dispersion involves mixing a metal salt or oxide with a substrate at a predetermined ratio. This ratio is determined by the amount of salt that is required for monolayer coverage on the surface of the substrate, assuming two-dimensional hexagonal close-packing. The BET surface area of the substrate is first measured. After the finely divided powders of the salt and substrate have been thoroughly mixed, it is heated at a temperature between the Tammann temperature and the melting point of the salt. This method is suitable for laboratory experiments. Another technique is incipient wetness impregnation, which is used in industrial scale for catalyst preparation. It involves preparing a solution of the salt to be dispersed. The solution is then mixed with the substrate, where it is absorbed by the substrate due to incipient wetness. After the substrate has imbibed the solution containing the salt into its pore structure, the sample needs to be heated to remove the solvent. Care needs to be taken when selecting solvents for use in this technique. First, the salt needs to be soluble in the solvent to a sufficient extent to allow enough salt to be dissolved in the volume of solution that can soaked by the pores of the substrate. Second, the solvent selected needs to be able to wet the surface of the substrate. Sorbents prepared by these two methods have been compared carefully (Padin and Yang, 2000). The sorbents prepared by incipient wetness were better. The equilibrium adsorption isotherms for C_2H_4 and C_2H_6 at 70°C on wet impregnated $AgNO_3/SiO_2$ are shown in Fig. 10.

The equilibrium isotherms of C_3H_6 over C_3H_8 at 70°C on $AgNO_3/SiO_2$ prepared by incipient wetness are shown in Fig. 11.

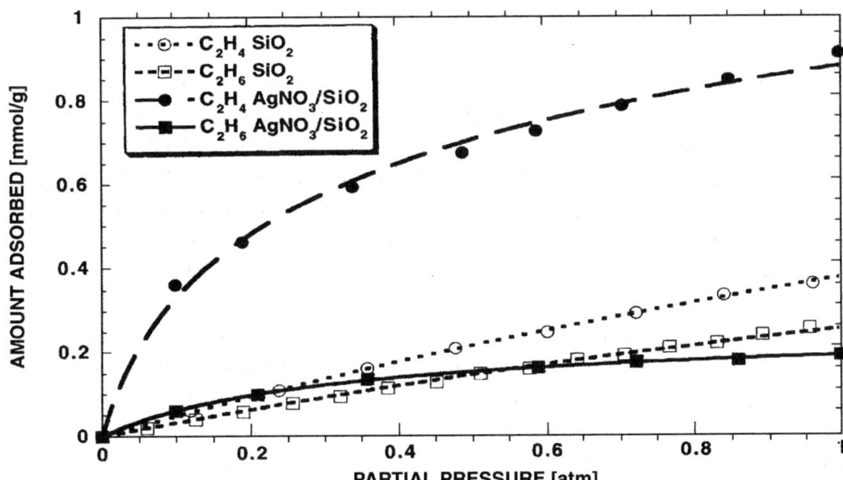

FIG. 10. Equilibrium isotherms of C_2H_4 over C_2H_6 on $AgNO_3/SiO_2$ (by incipient wetness) at 70°C (Padin and Yang, 2000).

From these results, it is seen that the sorbents have excellent selectivities and olefin capacities. The isotherms are also relatively linear. The linearity is desirable for cyclic processes such as PSA (Rege, Padin, and Yang, 1998). Diffusion rates and isotherm reversibilities have also been measured on these systems, and they were all highly suitable for PSA.

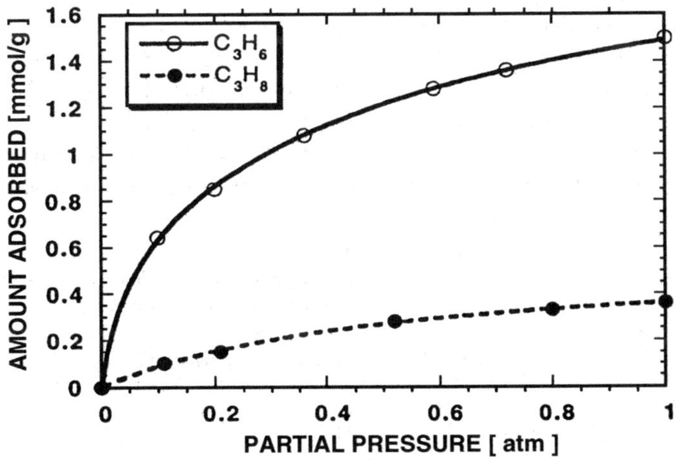

FIG. 11. Equilibrium isotherms of C_3H_6 over C_3H_8 on $AgNO_3/SiO_2$ (by incipient wetness) at 70°C (Padin and Yang, 2000).

B. Effects of Cation, Anion, and Substrate

An experimental as well as theoretical study has been performed in our laboratory on the effects of cations on adsorption (Huang, Padin, and Yang, 1999a). More specifically, a direct comparison of adsorption of CO and ethylene on AgCl and CuCl was made. An *ab initio* molecular orbital study using the effective core potential (ECP) was performed to determine the bond energies and the nature of the bonds between the adsorbates and adsorbents. Experimental results showed that both CO and C_2H_4 adsorbed more strongly on CuCl than on AgCl. However, CO adsorbed much more strongly on CuCl than on AgCl, whereas the difference in heats of adsorption was far less for C_2H_4 on these two sorbents. *Ab initio* molecular orbital calculations correctly predicted the trends. The natural bond orbital (NBO) theory was used to explain the results. The NBO results show that in the two-way π-complexation bonding, the d-π^* back-donation played a major role in determining the bonding for these systems (Huang, Padin, and Yang, 1999a).

The energies of adsorption calculated from molecular orbital theory (Huang, Padin, and Yang, 1999a) are shown in Table IX along with the experimental data. A good comparison between the theoretical values and experimental values is seen. Also, a perfect consistency with the geometry results can be obtained by comparing the differences in the energies of adsorption between the four adsorption systems. It is obvious from the table that CO adsorbed much more strongly on CuCl than on AgCl, with the heat of adsorption being about 7 kcal/mol larger, according to our calculation results. For the adsorption of C_2H_4 on these two adsorbates, however, the difference in the heat of adsorption is only about 4.5 kcal/mol. So it again shows that the difference in the adsorption of CO on CuCl and AgCl is much larger than that of C_2H_4 on these two adsorbates. Experiments and molecular orbital calculations were also performed for C_3H_6 on CuX and AgX (X = halide) (Huang, Padin, and Yang, 1999b).

Despite the fact that Cu^+ is better than Ag^+ in π-complexation, sorbents with Cu^+ are difficult to prepare due to the fact that its simple salts are not

TABLE IX
Energy of Adsorption (in kcal/mol) for MCl–C_2H_4 and MCl-CO Systems (M = Ag or Cu) (Huang, Padin, and Yang, 1999a)

Adsorbate	Adsorbent	Theoretical ΔH	Experimental ΔH (kcal/mol)
C_2H_4	AgCl	11.20	6.9
C_2H_4	CuCl	15.74	8.3
CO	AgCl	9.64	7.5
CO	CuCl	16.56	10.2

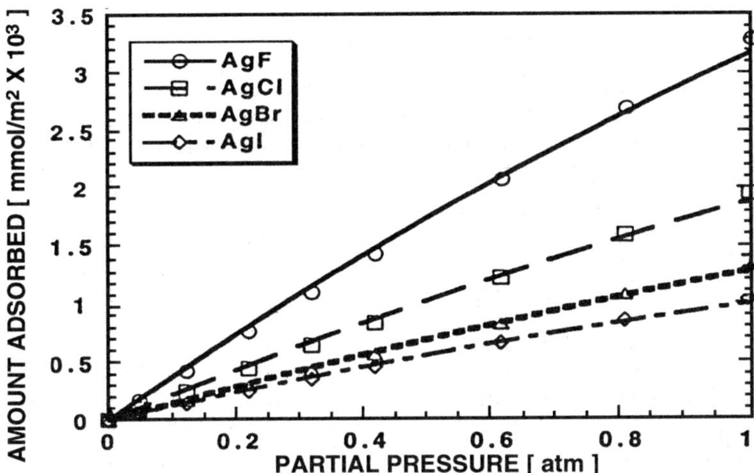

FIG. 12. Normalized C_3H_6 adsorption isotherm on AgX (X=F, Cl, Br, I) salts at 0°C (Huang, Padin, and Yang, 1999b).

water soluble. In addition, Cu^+ is not stable particularly upon exposure to moisture. More work is needed for preparing stable Cu^+ monolayer sorbents.

The effects of both cations and anions on π-complexation have been studied for the adsorption of C_2H_4 and C_3H_6 on CuX and AgX (X=F, Cl, Br, I), by experiment and molecular orbital theory (Huang, Padin, and Yang, 1999b). The following trends of anion and cation effects were obtained for the adsorption of C_2H_4 and C_3H_6 on the metal halides: $F^- > Cl^- > Br^- > I^-$ for anions and $Cu^+ > Ag^+$ for cations. These trends were in excellent agreement with the experimental results. In addition, the theoretical metal–olefin bond energies are in fair agreement with the experimental data. The anion effects are illustrated in Fig. 12.

The effects of substrates on π-complexation were studied by olefin adsorption on monolayer $AgNO_3$ supported on various substrates (Padin and Yang, 2000). The substrates selected were γ-Al_2O_3, SiO_2, and MCM-41. The following trend for olefin adsorption was observed for these substrates:

$$SiO_2 > \text{MCM-41} > \gamma\text{-}Al_2O_3. \qquad (20)$$

The silica surface (on both silica gel and MCM-41) provides a better substrate due to the lack of Lewis acid sites (unlike γ-Al_2O_3), and consequently the Ag atoms in these sorbents are more capable of forming π-complexation bonds with olefins. Although the effect of the physical characteristics of a substrate such as surface area and pore size would have on adsorption is clear, the effect of the electronic properties needs to be studied further.

TABLE X
SUMMARY OF THE NBO ANALYSIS OF π-COMPLEXATION BETWEEN MX AND C_2H_4 (HUANG, PADIN, AND YANG, 1999b)

	C → M interaction (σ donation)	M → C interaction (d-π^* back-donation)	Net change
	q_1[a]	q_2[a]	q_1+q_2
CuF–C_2H_4	0.047	–0.089	–0.042
CuCl–C_2H_4	0.052	–0.080	–0.028
CuBr–C_2H_4	0.042	–0.077	–0.035
CuI–C_2H_4	0.030	–0.072	–0.042
AgF–C_2H_4	0.081	–0.073	+0.008
AgCl–C_2H_4	0.058	–0.053	+0.004
AgBr–C_2H_4	0.047	–0.049	–0.002
AgI–C_2H_4	0.032	–0.044	–0.011

[a] q_1 is the amount of electron population increase on valence s orbitals of the metal and q_2 is the total amount of electron population decrease on valence d orbitals of the metal.

C. NATURE OF THE π-COMPLEXATION BOND

The nature of the metal–olefin bond was studied recently in our laboratory by analyzing the natural bond orbital (NBO) results (Huang, Padin, and Yang, 1999b). The main feature of the bonding can be seen from the population changes in the vacant outer-shell s orbital of the metal and those in the d shells of the metal upon adsorption. The NBO analysis, summarized in Tables X and XI, is generally in line with the traditional picture of Dewar (1951), and Chatt and Duncanson (1953) for metal–olefin complexation, i.e., it is dominated by the donation and back-donation contributions, as illustrated by Fig. 13.

An examination of Table XII shows that in all cases the M–C interaction is a dative bond, i.e., donation of electron charges from the π orbital of olefin to the vacant s orbital of metal and, simultaneously, back-donation of electron charges from the d orbitals of M to the π^* orbital of olefin (Fig. 12). This can be interpreted in more detail as follows. When the olefin molecule approaches M^+, some electronic charge is transferred from the C=C π orbital to the valence s orbital of M^+; at the same time, electrons in the filled d orbitals of metal are transferred to the symmetry-matched π^* orbital of olefin. It can be seen from Table XII that upon adsorption, the electron occupancies of the valence s orbitals of Cu and Ag always increase, whereas the total occupancy of their 4d or 5d orbitals always

TABLE XI
SUMMARY OF NBO ANALYSIS OF π-COMPLEXATION BETWEEN MX AND C_3H_6 (HUANG, PADIN, AND YANG, 1999b)

	$C \rightarrow M$ interaction (σ donation)	$M \rightarrow C$ interaction (d-π* back-donation)	Net change
	q_1[a]	q_2[a]	q_1+q_2
CuF–C_3H_6	0.046	−0.080	−0.034
CuCl–C_3H_6	0.051	−0.085	−0.034
CuBr–C_3H_6	0.040	−0.071	−0.031
CuI–C_3H_6	0.028	−0.067	−0.039
AgF–C_3H_6	0.081	−0.071	+0.010
AgCl–C_3H_6	0.060	−0.054	+0.006
AgBr–C_3H_6	0.046	−0.047	−0.001
AgI–C_3H_6	0.031	−0.043	−0.012

[a] q_1 is the amount of electron population increase on valence *s* orbitals of the metal, and q_2 is the total amount of electron population decrease on valence *d* orbitals of the metal.

decreases. Obviously this is caused by the donation and back-donation of electrons between metal and olefin, as stated before.

A comparison of the electron population changes in the *s* and *d* orbitals of M before and after adsorption shows that for the CuX–olefin complexes, the overall charge transfer is back-donation. The amount of back-donation is about double the amount of σ-donation. This indicates that the Cu–C bonds contain more metal *d* than metal *s* character and that the strength

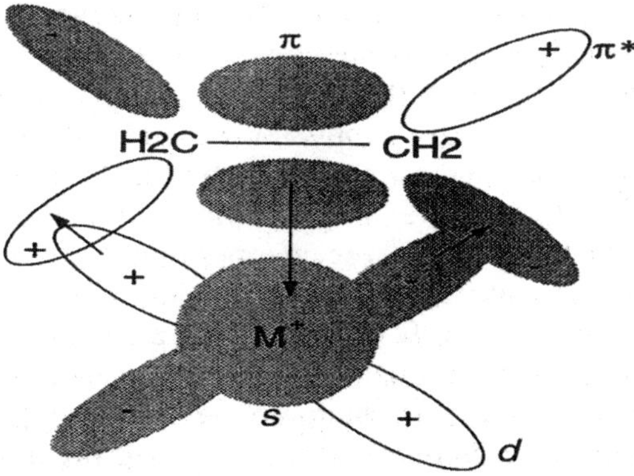

FIG. 13. Schematic representation of a metal–olefin complex showing σ-donation and d-π* back-donation.

TABLE XII
ELECTRON POPULATION CHANGES ON *d* ORBITALS OF Cu AND Ag AFTER C_2H_4 ADSORPTION (HUANG, PADIN, AND YANG, 1999b)

	dx_y	d_{xz}	d_{yz}	$d_{x^2-y^2}$	dz^2
CuF–C_2H_4	0	0	−0.0823	−0.0015	−0.0051
CuCl–C_2H_4	0	0	−0.0604	−0.0012	−0.0187
CuBr–C_2H_4	0	0	−0.0558	−0.0011	−0.0197
CuI–C_2H_4	0	0	−0.0505	−0.0010	−0.0202
AgF–C_2H_4	0	0	−0.0387	0	−0.0342
AgCl–C_2H_4	0	0	−0.0370	0	−0.0310
AgBr–C_2H_4	0	0	−0.0264	0	−0.0229
AgI–C_2H_4	0	0	−0.0239	0	−0.0201

of the covalent bonds depends mainly on the overlap of the metal *d* orbitals with the C hybrid orbitals. For the AgX–olefin complexes, quite differently, the back-donation is almost equal to the σ-donation, which means the σ-donation and back-donation play equally important roles in the bonding of Ag–C. A comparison of the net changes of the electron occupation on the two different metals before and after adsorption shows greater net electron occupation changes on Cu than on Ag upon olefin adsorption. The amount of change indicates the extent of interaction. This is consistent with the conclusion that CuX has a stronger interaction with olefin than AgX.

To gain further insight into the bonding between metal and olefin, Table XII lists the changes in electron populations of the five *d* orbitals of Cu and Ag upon ethylene adsorption.

As we can be seen clearly from the table, the interactions of the five *d* orbitals with ethylene are skewed. Orbitals d_{xy}, d_{xz} and $d_{x^2-y^2}$ have almost no contribution to the overlap with the π^* orbitals of the olefin because there is no or little change in their electron population upon ethylene adsorption. The main depopulation occurs in the d_{yz} and d_{z^2} orbitals. This is because the three inactive orbitals (d_{xy}, d_{xz} and $d_{x^2-y^2}$) are pointing in directions perpendicular to that of d_{yz} (in which plane the three member ring C–C–M lies), so there is little chance for them to overlap with the d_{yz} orbital. The depopulation in the d_{yz} orbital can be explained easily with the classic picture of π-complexation, shown in Fig. 13. However, the smaller amount of population decrease in the d_{z^2} orbitals is not expected. This phenomenon can be understood with the concept of *electron redistribution* that we proposed previously (Chen and Yang, 1996), illustrated in Fig. 14. As shown in Fig. 14, the dumbbell- and doughnut-shaped d_{z^2} orbitals are in the vicinity of the spatial directions of

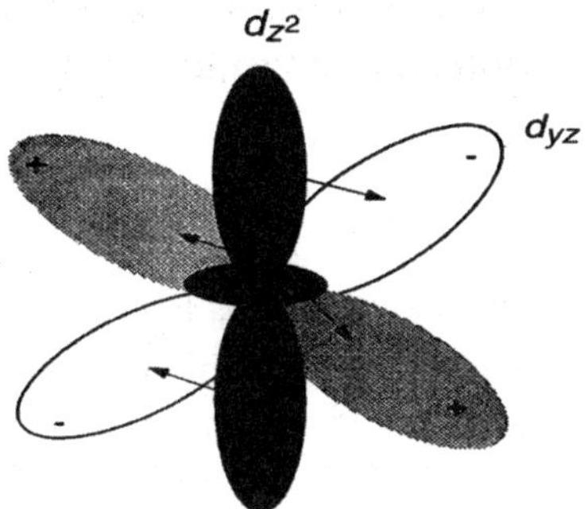

FIG. 14. Schematic representation of electron redistribution from metal d_{z^2} to d_{yz}.

the d_{yz} orbitals, and they can overlap to some extent with the d_{yz} orbitals. This result indicates that there is a considerable electron redistribution between the d_{yz} and d_{z^2} orbitals during the metal–olefin bonding. Obviously, electron redistribution from the d_{z^2} to the d_{yz} orbitals helped enhance the d-π^* back-donation.

D. OLEFIN–DIENE SEPARATION AND PURIFICATION, AROMATIC AND ALIPHATICS SEPARATION, AND ACETYLENE SEPARATION

Sorbents and separations based on π-complexation have also found use in other possible applications. Ag^+ ion–exchanged X or Y zeolites showed an excellent capability for purification of olefins by removing trace amounts of corresponding dienes. This has been demonstrated for the butadiene/butene system (Padin, Yang, and Munson, 1999).

For the separations of aromatics from aliphatics, sorbents based on π-complexation have also been studied (Takahashi, Yang, and Yang, 2000). The benzene–cyclohexane pair was used as a model compound. A number of transition metal salts were dispersed in monolayer amounts on a high-surface-area substrate. $PdCl_2$ or $AgNO_3$ dispersed on SiO_2 gel exhibited high equilibrium adsorption ratios of benzene over cyclohexane. $PdCl_2$ loading of 0.88 g/g-SiO_2 showed the best benzene–cyclohexane ratio of 3.2.

For selective acetylene adsorption from other hydrocarbons (e.g., ethylene and ethane), $NiCl_2$ supported on alumina or silica can form reversible π-complexation bonds with acetylene but not olefins. Pure component acetylene–ethylene ratios of up to 3 were obtained (Kodde *et al.*, 2000). The bonding between acetylene and $NiCl_2$ is reasonably understood (Huang and Yang, 1999).

VII. Other Sorbents and Their Unique Adsorption Properties: Carbon Nanotubes, Heteropoly Compounds, and Pillared Clays

A. Carbon Nanotubes

Carbon nanotubes are derivatives of the C_{60} buckyballs. They are formed by graphite (or graphene) sheets rolled up into tubes, generally in the range of 2 to 10 nm in diameter and 200 to 500 nm in length. The multiwall nanotubes were described by Iijima (1991). Single-wall nanotubes were discovered in 1993 (Iijima and Ichihashi, 1993; Bethune *et al.,* 1993). Since their discovery, these materials have attracted intense interests due to their potential in applications in a variety of nanotechnologies (Dresselhaus, Dresselhaus, and Eklund, 1995), such as molecular electronics and scanning probe microscope tips. Of particular interest in adsorption is hydrogen storage.

Much excitement has arisen on recent reports of promising results on carbon nanotubes for hydrogen storage (Dillon *et al.,* 1997; Chambers *et al.,* 1998; Chen, Lin, and Tan, 1999; Liu *et al.,* 1999). High hydrogen adsorption capacities were reported for various carbon nanotubes.Dillon *et al.* (1997) reported that single-wall nanotubes could potentially store up to 5 to 10% by weight of hydrogen at 273 K and 300 torr based on estimates from temperature-programmed desorption data.Chambers *et al.* (1998), using a volumetric system, reported up to 56% by weight of hydrogen storage by multiwall nanotubes at 120 atm and 298 K. More recently, interesting results were reported for using alkali-doped carbon nanotubes for hydrogen storage (Chen, Lin, and Tan, 1999). It was reported that Li- and K-doped carbon nanotubes adsorbed, respectively, 20 wt.% and 14 wt.% of hydrogen at 1 atm and mild temperatures (200 to 400°C for Li-doped and near room temperature for K-doped nanotubes). Lower but still substantial amounts of hydrogen adsorption were also reported for alkali-doped graphite (Chen, Lin, and Tan, 1999). Most recently, Liu *et al.* (1999) reported about 4% by weight of hydrogen adsorption in single-wall nanotubes, at 100 atm and room temperature. It should be noted that the goal for hydrogen storage set by the Department of Energy is 5% by weight for onboard automotive applications.

Needless to say, hydrogen storage with carbon nanotubes has tremendous potentials, yet it is in a highly confusing state. It is difficult, if not impossible, to reproduce data for several reasons. First, carbon nanotubes can be prepared in a number of ways, and it is difficult to reproduce the nanotubes prepared in another laboratory even if the same technique is followed. Second, leakage of hydrogen at high pressures is a notorious problem in the laboratory. In the measurement using a typical volumetric system, the signal for hydrogen adsorption is typically a few psi, which can be easily the amount of leakage.

As mentioned, Chen *et al.* (1999) reported recently in *Science* the most promising results of hydrogen storage in alkali-doped nanotubes at 1 atm pressure. The nanotubes were multiwall nanotubes prepared by catalytic decomposition of methane on $Ni_{0.4}Mg_{0.6}O$. The amount of hydrogen storage was measured with a microbalance. Both the nanotubes and the measurement were easily reproducible. This author has revisited the problem (Yang, 2000). It was found that most of the weight gain was caused by the moisture impurity contained in the hydrogen gas cylinder. Gases supplied in cylinders are frequently contaminated by moisture, and this is particularly the case with hydrogen.

B. Heteropoly Compounds

Heteropoly compounds are a class of crystalline materials with unique structures (Pope, 1983). They can be represented by $H_3XM_{12}O_{40}$. The crystalline structure of the anion $XM_{12}O_{40}$ belongs to the Keggin structure. In this structure, 12 MO_6 octahedra surround a central XO_4 tetrahedron, where M is usually W or Mo and X can be P, As, Si, Ge, B, and others. Although the structures of the heteropoly anions (i.e., the Keggin structure) are well defined and stable (termed primary structure), the structures by which the Keggin units are linked together (termed secondary structure) are less understood. Water is usually the linkage molecule.

$H_3PW_{12}O_{40}$ is one of the most common heteropoly compounds. A distinct X-ray diffraction pattern is seen for $H_3PW_{12}O_{40}.6H_2O$, where the Keggin units are linked by $H^+(H_2O)_2$ bridges, resulting in a body-centered cubic structure and hence the X-ray diffraction pattern. The water molecules can be easily replaced by a number of polar molecules such as alcohols and amines (Misono, 1987).

The unique ability of the heteropoly compound to selectively bind NO was recently found in the author's laboratory (Yang and Chen, 1994; Chen and Yang, 1995). The water linkages in $H_3PW_{12}O_{40}.6H_2O$ can be readily replaced by NO linkages at 50–230°C at low NO concentrations (i.e., under flue-gas conditions) to form $H_3PW_{12}O_{40}.3NO$ (Chen and Yang, 1995). More interestingly, a substantial fraction of the absorbed NO is decomposed into N_2 upon rapid heating of the NO-linked compound (Yang and Chen, 1994; Chen and Yang, 1995). NO is absorbed selectively over other molecules in the flue gas (i.e., CO_2, SO_2, H_2O). These are unique properties of the heteropoly compound. The NO-linked compound has been characterized in detail (Chen and Yang, 1995). This work has been followed by a number of groups, and excellent work has been done by McCormick and coworkers (Herring, McCormick, and Boonrueng, 2000).

C. Pillared Clays

Pillared interlayered clays (PILCs), or pillared clays, are two-dimensional layer materials prepared by exchanging the charge-compensating cations (e.g., Na^+, K^+, and Ca^{2+}) between the swelling phyllosilicate clay layers with large inorganic hydroxy cations, which are polymeric or oligomeric hydroxy metal cations formed by hydrolysis of metal salts. Upon heating, the metal hydroxy cations undergo dehydration and dehydroxylation, forming stable metal oxide clusters that act as pillars, keeping the silicate layers separated and creating interlayer spacing (gallery spaces) of molecualr dimensions. A number of pillared clays can be made easily, such as pillared clays with the following metal oxide pillars: Al, Zr, Cr, Fe, Ti, Si, and Ni.

The cations on the as-made pillared clays are protons. The ion-exchange capacity of the original clay is preserved in the final pillared clay. Hence, PILCs can have very large ion-exchange capacities, for example, 140 meq/g for the Arizona montmorillonite. The high ion-exchange capacities of PILCs are potentially useful for ion exchange applications.

The adsorption propoerties of PILCs and some ion-exchanged PILCs have been studied by a number of groups, including our own. Yang and Baksh (1991) suggested that the pore dimensions in PILCs are not limited by the interlayer spacing (that is determined by X-ray diffraction), but is limited by interpillar spacing. Moreover, the interlayer spacing can be tailored by controlling the number density of pillars that are inserted during the ion-exchange step in the synthesis. The pore-size distributions of the PILCs are actually bimodal: a major portion of pore volume is in pores near 0.6-0.8 nm and a small fraction of pore volume near 0.4 nm (Hutson, Gualdoni, and Yang, 1998; Gil and Grange, 1996; Gandia and Vicente, 2000). By ion exchange with cations of different sizes, the pore dimensions can be furthered tailored (Hutson, Gualdoni, and Yang, 1998). Hence, PILCs offer versatility in both pore-size distribution (by tailoring) as well as surface chemistry (by using different pillars and also by ion exchange). These make PILCs potentially useful as sorbents. However, this promise remains to be fulfilled.

ACKNOWLEDGMENTS

I am grateful to my past and current students and co-workers with whom I have had so much pleasure in learning. Ms. Peggy Kuch's assistance in formatting the manuscript is sincerely acknowledged. This work was supported by the National Science Foundation.

REFERENCES

Adamson, A. W., "Physical Chemistry of Surfaces." 3rd ed., Wiley, New York, 1976.
Alcaniz-Monge, J., de la Casa-Lillo, M. A., Cazorla-Amoros, D., and Linares-Solano, A., *Carbon* **35,** 291 (1997).
Antochsuk, V., and Jaroniec, M., *Chem. Mater.* **12,** 6271 (2000).
Barrer, R. M., "Zeolites and Clay Minerals." Academic Press, New York (1978).
Barrer, R. M., and Gibbons, R. M., *Trans. Faraday Soc.* **61,** 948 (1965).
Barrer, R. M., and Stuart, W. I., *Proc. Roy. Soc.* **A249,** 464 (1959).
Barton, T. J., Bull, L. M., Klemperer, W. G., Loy, D. A., McEnaney, B., Misono, M., Monson, P. A., Pez, G., Sherer, G. W., Vartuli, J. A., and Yaghi, O. M., *Chem. Mater.* **11,** 2633 (1999).
Beck, J. S., Vartuli, J. C., Roth, W. J., Leonowicz, M. E., Kresge, C. T., Schmitt, K. D., Chu, C. T.-W., Olsen, D. H., Sheppard, E. W., McCullen, S. B., Higgins, J. B., and Schlenker, J. L., *J. Am. Chem. Soc.* **114,** 10834 (1992).
Bethune, D. S., Kiang, C. -H., de Vries, M. S., Gorman, G., Savoy, R., Vasquez, J., and Beyers, R., *Nature* **363,** 605 (1993).
Blytas, G. C., Separation of unsaturates by complexing with nonaqueous solutions of cuprous salts, *In* "Separation and Purification Technology" (N. N. Li, and J. M., Clao, Eds.), Chap. 2. Marcel Dekker, New York, 1992.
Breck, D. W., "Zeolite Molecular Sieves," Wiley, New York (1974).
Brinker, C. J., and Sherer, G. W., "Sol-Gel Science." Academic Press, New York, 1990.
Carrasco-Marin, F., Mueden, A., Centeno, T. A., Stoeckli, F., and Moreno-Castilla, C. J., *Chem. Soc. Faraday Trans.* **93,** 2211 (1997).
Chambers, A., Park, C., Baker, R. T. K., and Rodriguez, N. M., *J. Phys. Chem. B* **102,** 4253 (1998).
Chao, C. C., U.S. Patent 4,859,217 (1989).
Chao, C., Sherman, J. D., Mullhaupt, T. J., and Bollinger, C. M., U.S. Patent 5,174,979 (1992).
Chatt, J., and Duncanson, J. A., *J. Chem. Soc.* 2939 (1953).
Chen, P., Wu, X., Lin, J., and Tan, K. L., *Science* **285,** 91 (1999).
Chen, N., and Yang, R. T., *J. Catal.* **157,** 76 (1995).
Chen, N., and Yang, R. T., *Ind. Eng. Chem. Res.* **35,** 4020 (1996).
Cheng, L. S., and Yang, R. T., *Chem. Eng. Sci.* **49,** 2599 (1994).
Cheng, L. S., and Yang, R. T., *Adsorption* **1,** 61 (1995).
Coe, C. G., *in* "Access in Nanoporous Material," (T. J. Pinnavaia and M. F. Thorpe, Eds.), p. 213. Plenum Press, New York, 1995.
Cracknell, R. F., Gubbins, K. E., Maddox, M., and Nicholson, D., *Acc. Chem. Res.* **28,** 281 (1995).
Dewar, M. J. S., *Bull. Soc. Chim., Fr.* **18,** C71 (1951).
Dillon, A. C., Jones, K. M., Bekkedahl, T. A., Kiang, C. -H., Bethune, D. S., and Heben, M. J., *Nature* **386,** 377 (1997).
Doong, S. J., and Yang, R. T., *AIChE Symp. Ser.* **83,** 87, AIChE, New York (1987).
Dresselhaus, M. S., Dresselhaus, G., and Eklund, P. C., "Science of Fullerenes and Nanotubes," Academic Press, New York, 1995.
Eldridge, R. B., *Ind. Eng. Chem. Res.* **32,** 2208 (1993).
Feng, X., Fryxell, G. E., Wang, L. -Q., Kim, A. Y., Liu, J., and Kemmer, K. M., *Science* **276,** 923 (1997).
Ghosh, T. K., Lin, H. D., and Hines, A. L., *Ind. Eng. Chem. Res.* **32,** 2390 (1993).
Gil, A., Gandia, L. M., and Vicente, M. A., *Catal. Rev. Sci. Eng.* **42,** 145 (2000).
Gil, A., and Grange, P., *Colloids and Surfaces* **113,** 39 (1996).

Gilliland, E. R., Bliss, H. L., and Kip, C. E., *J. Am. Chem. Soc.* **63,** 2088 (1941).
Gregg, S. J., and Sing, K. S.W., "Adsorption, Surface Area and Porosity." 2nd ed., Academic Press, New York, 1982.
Hall, C. R., and Holmes, R. J., *Carbon* **31,** 881 (1993).
Henrich, V. E., and Cox, P. A., "The Surface Science of Metal Oxides." Cambridge University Press, Cambridge, UK, 1994.
Herring, A. M., McCormick, R. L., and Boonrueng, S. K., *J. Phys. Chem. B* **19,** 4653 (2000).
Hirai, H., Hara, S., and Komiyama, M., *Angew. Makromol. Chem.* **130,** 207 (1985).
Hirai, H., Kurima, K., Wada, K., and Komiyama, M., *Chem. Lett. (Japan)* **1513** (1985).
Ho, W. S., Doyle., G., Savage, D. W., and Pruett, R. L., *Ind. Eng. Chem. Res.* **27,** 334 (1988).
Horvath, G., and Kawazoe, K., *J. Chem. Eng. Japan* **16,** 470 (1983).
Huang, H. Y., Padin, J., and Yang, R. T., *Ind. Eng. Chem. Res.* **38,** 2720 (1999a).
Huang, H. Y., Padin, J., and Yang, R. T., *J. Phys. Chem. B.* **103,** 3206 (1999b).
Huang, H. Y., and Yang, R. T., *Langmuir* **15,** 7647 (1999).
Huggahalli, M., and Fair, J. R., *Ind. Eng. Chem. Res.* **35,** 2071 (1996).
Humphrey, J. L., and Keller, II, G. E., "Separation Process Technology." McGraw-Hill, New York, 1997.
Huo, Q., Mragolese, D. I., and Stucky, G. D., *Chem. Mater.* **8,** 1147 (1996).
Hutson, N. D., Gualdoni, D. J., and Yang, R. T., *Chem. Mater.* **10,** 3707 (1998).
Hutson, N. D., Rege, S. U., and Yang, R. T., **45,** 724 (1999).
Iijima, S., *Nature* **354,** 56 (1991).
Iijima, S., and Ichihashi, T., *Nature* **363,** 603 (1993).
Iler, R. K., "The Chemistry of Silica." Wiley, New York, 1979.
Israelachilli, J., "Intermolecular and Surface Forces." 2nd ed., Academic Press, New York, 1992.
Izumi, J., "Mitsubishi VOC Recovery Process." Mitsubishi Heavy Industries, Ltd., 1996. Cited from (Tanev and Pinnavaia, 1995).
Jankowska, H., Swiatkowski, A., and Choma, J., "Active Carbon." Ellis Harwood, New York, 1991.
Jaroniec, C. P., Kruk, M., Jaroniec, M., and Sayari, A., *J. Phys. Chem. B.* **102,** 5503 (1998).
Jarvelin, H., and Fair, J. R., *Ind. Eng. Chem. Res.* **32,** 2201 (1993).
Jiang, S., Zollweg, J. A., and Gubbins, K. E., *J. Phys. Chem.* **98,** 5709 (1994).
Jones, R. W., "Fundamental Principles of Sol-Gel Technology." The Institute of Metals, London, 1989.
Kaman, N. K., Anderson, M. T., and Brinker, C. J., *Chem. Mater.* **8,** 1682 (1996).
Kaneko, K., *Carbon* **38,** 287 (2000).
Kaneko, Y., Ohbu, K., Uekawa, N., Fugie, K., and Kaneko, K., *Langmuir* **11,** 708 (1995).
Kaneko, K., Yang, C. M., Ohkubo, T., Kimura, T., Iiyama, T., and Touhara, H., *in* "Adsorption Science and Technology" (D. D. Do, Ed.), p.1. World Scientific Publishers, Singapore, 2000.
Keller, G. E., Marcinkowsky, A. E., Verma, S. K., and Wiliamson, K. D., Olefin recovery and purification via silver complexation, "Separation and Purification Technology" (N. N. Li, and J. M. Calo, Eds.), p.59. Marcel Dekker, New York, 1992.
King, C. J., Separation processes based on reversible chemical complexation, *in* "Handbook of Separation Process Technology" (R. W. Rousseau, Ed.), Chap. 15. Wiley, New York, 1987.
Kodde, A. J., Padin, J., van der Meer, P. J., Mittelmeijer-Hazeleger, M. C., Bliek, A., and Yang, R. T., *Ind. Eng. Chem. Res* **39,** 3108 (2000).
Kulvaranon, S., Findley, M. E., and Liapis, A. I., *Ind. Eng. Chem. Res.* **29,** 106 (1990).
Kumar, R., Kratz, W. C., Guro, D. E., and Golden, T. C., A new process for the production of high purity carbon monoxide and hydrogen, Presented at the Int. Symp. On Separation Technology, Univ. Of Antwerp, Belgium (Aug. 22–27, 1993).
Lamari, M., Aoufi, A., and Malbrunot, P., *AIChE J.* **46,** 632 (2000).

Leon y Leon, C. A., and Radovic, L. R., *in* "Chemistry and Physics of Carbon," (P. A. Thrower, Ed.), Vol. 24, p. 213. Dekker, New York, 1992.
Liu, C., Fan, Y. Y., Liu, M., Cong, H. T., Cheng, H. M., and Dresselhaus, M. S., *Science* **286,** 1127 (1999).
Long, R. B., Separation of unsaturates by complexing with solid copper salts, *in* "Recent Developments in Separation Science" (N. N. Li, Ed.), Vol. 1, p.35. CRC Press, Cleveland, 1972.
Masel, R. I., "Principles of Adsorption and Reaction on Solid Surfaces." Wiley, New York, 1996.
Matranga, K. R., Myers, A. L., and Glandt, E. D., *Chem. Eng. Sci.* **47,** 1569 (1992).
McKee, D. W., U.S. Patent 3,140,933 (1964).
Mellot, C., and Lignieres, J., *in* "Physical Adsorption: Experiment, Theory and Applications" (J. Fraissard, and C. W. Conner. Eds.), Kluwer Academic, Boston, 1997.
Milton, R. M., U.S. Patent 2,882,243, To Union Carbide Corporation, (1959).
Milton, R. M., U.S. Patent 2,882,244, To Union Carbide Corporation, (1959).
Misono, M., *Catal. Rev. Sci. Eng.* **29,** 269 (1987).
Moller, K., and Bein, T., *Chem. Mater.* **10,** 2950 (1998).
Mota, J. P. B., *AIChE J.* **45,** 986 (1999).
Padin, J., and Yang, R. T., *Chem. Eng. Sci.* **55,** 2607 (2000).
Padin, J., Yang, R. T., and Munson, C. L., *Ind. Eng. Chem. Res.* **38,** 3614 (1999).
Pope, M. T., "Heteropoly and Oxopoly Oxometalates," Springer-Verlag, New York, 1983.
Quinn, H. W., Hydrocarbon separations with silver (I) systems, *in* "Progress in Separation and Purification" (Perry, Ed.), Vol. 4, p.133. Interscience, New York, 1971.
Razmus, D. M., and Hall, C. K., *AIChE J.* **37,** 769 (1991).
Rege, S. U., Padin, J., and Yang, R. T., *AIChE J.* **44,** 799 (1998).
Rege, S. U., and Yang, R. T., *AIChE J.* **46,** 734 (2000).
Rigby, M., Smith, E. B., Wakeham, W. A., and Maitland, G. C., "The Forces Between Molecules." Oxford University Press, Fairlaun, NJ 1986.
Rodriguez-Reinoso, F., Molina-Sabio, M., and Munecas, M. A., *J. Phys. Chem.* **96,** 2707 (1992).
Ross, S., and Olivier, J. P., "On Physical Adsorption." Wiley, New York, 1964.
Rouquerol, F., Rouquerol, J., and Sing, K., "Adsorption by Powders and Porous Solids." Academic Press, San Diego, 1999.
Russel, B. P., and LeVan, M. D., *Ind. Eng. Chem. Res.* **36,** 2380 (1997).
Ruthven, D. M., "Principles of Adsorption and Adsorption Processes." Wiley, New York, 1984.
Safarik, D. J., and Eldridge, R. B., *Ind. Eng. Chem. Res.* **37,** 2571 (1998).
Saito, A., and Foley, H. C., *AIChE J.* **37,** 429 (1991).
Salame, I. I., and Bandosz, T. J., *J. Coll. Interf. Sci.* **210,** 367 (1999).
Sayari, A., Liu, P., Kruk, M., and Jaroniec, M., *Chem. Mater.* **9,** 2499 (1997).
Shepelev, Y. F., Anderson, A. A., and Smolin, Y. I., *Zeolites* **10,** 61 (1990).
Steele, W. A., "The Interaction of Gases with Solid Surfaces." Pergamon Press, New York, 1974.
Sun, T., and Ying, J. Y., *Nature* **389,** 704 (1997).
Suzuki, M., *Carbon* **32,** 577 (1994).
Szostak, R., "Molecular Sieves." 2nd ed., Blackie Academic & Professional, New York 1998.
Takahashi, A., Yang, F. H., and Yang, R. T., *Ind. Eng. Chem. Res.* **39,** (2000).
Talu, O., *in* "Fundamentals of Adsorption" (M. Suzuki, Ed.), p. 655. Elsevier, Amsterdam, 1993.
Tanev, P. T., and Pinnavaia, T. J., *Science* **267,** 865 (1995).
Teraoka, Y., Fukunaga, Y., Setaguchi, Y. M., Moriguchi, I., Kagawa, S., Tomonago, N., Yasutake, A., and Isumi, J., *in "Adsorption Science and Technology."* (D. D. Do, Ed.), p. 603. World Scientific Publishers, Singapore, 2000.
Wu, S., Han, S., Cho, S. H., Kim, J. N., and Yang, R. T., *Ind. Eng. Chem. Res.* **36,** 2749 (1999).

Xie, Y. C., and Tang, Y. Q., *Adv. Cat.* **37,** 1 (1990).
Yang, R. T., "Gas Separation by Adsorption Processes." Butterworth, Boston, 1987; Imperial College Press, London, 1997.
Yang, R. T., *Carbon* **38,** 623 (2000).
Yang, R. T., and Baksh, M. S. A., *AIChE J.* **37,** 679 (1991).
Yang, R. T., and Chen, N., *Ind. Eng. Chem. Res.* **33,** 825 (1994).
Yang, R. T., and Hutson, N. D., Lithium-based zeolites containing silver and copper and use thereof for selective adsorption, U.S. Patent 60/114317 (December, 1998).
Yang, R. T., and Kikkinides, E. S., *AIChE Journal,* **41,** 509 (1995).
Yang, F. H., and Yang, R. T., *Carbon,* in press (2001).
Ying, J. Y., Mehnert, C. P., and Wong, M. S., *Angew. Chem. Int. Ed.* **38,** 56 (1999).
Young, D. M., and Crowell, A. D., "Physical Adsorption of Gases." Butterworth, London, 1962.
Zhao, D., Feng, J., Huo, Q., Melosh, N., Frederickson, G. H., Chmelka, B. F., and Stucky, G. D., *Science* **279,** 548 (1998).
Zhao, X. S., Lue, G. Q., and Hu, X., *Chem. Comm.* **1391** (1999).
Zhao, X. S., Lu, G. Q., and Millar, G. J., *Ind. Eng. Chem. Res.* **35,** 2075 (1996).

NANOPHASE CERAMICS: THE FUTURE ORTHOPEDIC AND DENTAL IMPLANT MATERIAL

Thomas J. Webster

Department of Biomedical Engineering, Purdue University, West Lafayette, Indiana 47907

I. Introduction 126
II. Mechanical Properties of Bone 128
III. Bone Physiology 128
A. Microarchitecture 128
B. Structural Organization of the Bone Microarchitecture 131
C. Chemical Composition of the Bone Matrix 131
D. Cells of the Bone Tissue 136
E. Bone Remodeling 139
IV. The Tissue–Implant Interface 140
A. Wound-Healing Response of Bone 141
B. Protein Interactions with Biomaterial Surfaces 141
C. Protein-Mediated Cell Adhesion on Biomaterial Surfaces 143
V. Materials Currently Used as Orthopedic and Dental Implants 145
A. Novel Surface Modifications of Conventional Orthopedic and Dental Implants 147
VI. Next Generation of Orthopedic and Dental Implants: Nanophase Ceramics 148
A. Surface Properties of Nanophase Ceramics for Enhanced Orthopedic and Dental Implant Efficacy 149
B. Mechanical Properties of Nanophase Ceramics for Enhanced Orthopedic/Dental Implant Efficacy 156
VII. Conclusions 159
References 160

Traditional materials utilized for orthopedic and dental applications have been selected based on their mechanical properties and ability to remain inert in vivo; this selection process has provided materials that satifisfy physiological loading conditions but do not

ADVANCES IN CHEMICAL ENGINEERING, VOL. 27
0065-2377/01 $35.00

duplicate the mechanical, chemical, and architectural properties of bone. The less than optimal surface properties of conventional materials have resulted in clinical complications that necessitate surgical removal of many such failed bone implants due to insufficient bonding to juxtaposed bone. Sufficient bonding of an implant to juxtaposed bone (i.e., osseointegration) is needed to minimize motion-induced damage to surrounding tissues and support physiological loading conditions, criteria crucial for implant success. Insufficient osseointegration can be caused by biomaterial surface properties that do not support new bone synthesis and/or mechanical properties that do not match those of surrounding bone; mismatch of mechanical properties between an implant and surrounding bone may lead to stress and strain imbalances that cause implant loosening and eventual failure. Clearly, the next generation of biomaterials for orthopedic and dental implant applications must possess both biocompatible surface properties that promote bonding of juxtaposed bone and mechanical properties similar to those of physiological bone. Due to unique surface and mechanical properties, as well as the ability to simulate the three-dimensional architecture of physiological bone, one possible consideration for the next generation of orthopedic and dental implants with improved efficacy are nanophase materials. This chapter presents reports of the design, synthesis, and evaluation of nanophase materials for increased orthopedic and dental implant efficacy.

I. Introduction

An estimated 11 million people in the United States have received at least one medical implant device; orthopedic implants (including fracture, fixation, and artificial joint devices) accounted for 51.3% of these implants in 1992 (Praemer *et al.*, 1992). Among joint-replacement procedures, hip and knee surgeries represented 90% of the total and in 1988 were performed 310,000 times in the United States alone (Praemer *et al.*, 1992). Implanting an orthopedic or dental implant can be a costly procedure (due to surgery, hospital-provided care, physical therapy costs, and recuperation time) and involves considerable patient discomfort. Both patient discomfort and cost can increase if surgical revision becomes necessary after implantation when an orthopedic or dental implant is rejected by host tissue, is insufficiently integrated into juxtaposed bone, and/or fails under physiological loading conditions. Careful design of implants with improved properties (such as biocompatibility, ability to enhance integration into surrounding bone, and mechanical properties similar to physiological bone) could increase biomaterial success rate and, therefore, decrease patient discomfort as well as surgical costs associated with device retrieval and implantation of another orthopedic and dental prostheses.

Successful biomaterials must integrate into surrounding tissue by eliciting timely and desirable responses from surrounding cells. Orthopedic and dental biomaterials of the future must promote swift deposition of new bone on the surface of implants and support bonding of juxtaposed bone (that is, osseointegrate) to stabilize prostheses *in situ* and thus minimize motion-induced damage to surrounding tissues. In addition, and equally as important, biomaterials of the future must possess mechanical properties similar to those of surrounding bone; mismatch between mechanical properties of implants and surrounding bone could cause imbalances in stress and strain distribution, thus, leading to bone resorption and eventual implant loosening or failure (Kaplan *et al.,* 1994; Brunski, 1991; Lehman *et al.,* 1994).

Conventional orthopedic and dental materials (such as commercially pure titanium, Ti–6Al–4V, and Co–Cr–Mo alloys) meet requirements for mechanical loads in the physiological range, but their less than optimal surface properties (leading to insufficient osseointegration) have resulted in clinical complications and necessitated surgical removal of many such failed bone implants (Kaplan *et al.,* 1994). In contrast, ceramics, which have exceptional biocompatibility and surface properties with bone cells and tissues but are brittle under loading, have experienced limited use in biomedical applications. The next generation of biomaterials for orthopedic and dental implant applications must possess both biocompatible surface properties that promote bonding of juxtaposed bone and mechanical properties similar to those of physiological bone.

One possible consideration for the next generation of orthopedic and dental implants with improved efficacy are nanophase materials. Nanophase materials are new material formulations that possess grains composed of the same atoms, but the atoms are fewer (less than tens of thousands) and smaller (less than 100 nm in diameter) than in conventional materials (which contain several billions of atoms and grain sizes of micrometers to millimeters) (Siegel, 1996). Control of the size of the constituent clusters and the manner in which these clusters are assembled in nanostructures has produced new materials with unique, custom-made mechanical, electrical, chemical, magnetic, and optical structures and properties (Siegel, 1996). Despite their great promise, investigations of nanophase materials as orthopedic and dental implants have been close to nonexsistent.

This chapter will discuss, in detail, properties of bone and of nanophase ceramics that promise increased orthopedic and dental implant efficacy.

Limited reports in the literature that discuss the design, synthesis, and evaluation of nanophase ceramics for orthopedic and dental implants will also be discussed.

II. Mechanical Properties of Bone

Bone is a well-organized tissue whose primary function is to support mechanical loading and protect vital organs in the body (Martin and Burr, 1989). A unique living tissue, bone possesses the ability to regenerate itself and adapt its geometry to accommodate local stress and strain from the surrounding physiological milieu. In the course of normal, daily activities, mechanical loads are applied to bone; for example, human jaws sustain 0.25 kN during chewing, and human hip joints are exposed to 3 to 5 kN and 0.75 kN during walking and standing, respectively (Kaplan *et al.,* 1994; Park and Lakes, 1992). Mechanical properties of bone change with architecture (i.e., cortical or trabecular), porosity (for example, 30% and 50 to 90% porosity for cortical and trabecular bone, respectively), anatomical differences (for example, the modulus of elasticity varies by 10% in bone from the human hip and tibia), and age (e.g., the modulus of elasticity varies by up to 20% in bone from 20- to 90-year-old humans) (Kaplan *et al.,* 1994; Fung, 1993).

The mechanical properties of cortical bone (specifically, the femur, tibia, humerus, and radius) of various species (specifically, horse, cattle, pig, and human) in tension, compression, and torsion are listed in Table I. It should be noted, for example, that human femur tensile strength (namely, 124 MPa) (Yamada, 1970) is in the same order of magnitude to that of cast iron (170 MPa) (Beer and Johnston, 1981) but, surprisingly, low in weight (Kaplan *et al.,* 1994; Fung, 1993). These unique properties of bone are a direct consequence of the synergy of its molecular, cellular, and tissue arrangement.

III. Bone Physiology

A. Microarchitecture

At the microscopic level, bone consists of two structures: woven and lamellar (Fig. 1a). Woven bone (with an average mineral grain size of 10 to 50 nm) is the immature, or primitive, form of bone and is normally found in the metaphyseal region of growing bone as well as in fracture callus (Kaplan *et al.,* 1994; Park and Lakes, 1992). Woven bone is coarse-fibered and

TABLE I
MECHANICAL PROPERTIES OF CORTICAL BONE IN TENSION, COMPRESSION, AND TORSION[a]

Bone type	Horses	Cattle	Pigs	Humans (20–39 years)
		Ultimate tensile strength (MPa)		
Femur	121 ± 1.8	113 ± 2.1	88 ± 1.5	124 ± 1.1
Tibia	113	132 ± 2.8	108 ± 3.9	174 ± 1.2
Humerus	102 ± 1.3	101 ± 0.7	88 ± 7.3	125 ± 0.8
Radius	120	135 ± 1.6	100 ± 3.4	152 ± 1.4
		Ultimate percent elongation (%)		
Femur	0.75 ± 0.01	0.88 ± 0.02	0.68 ± 0.01	1.41
Tibia	0.70	0.78 ± 0.01	0.76 ± 0.03	1.50
Humerus	0.65 ± 0.01	0.76 ± 0.01	0.70 ± 0.03	1.43
Radius	0.71	0.79 ± 0.01	0.73 ± 0.03	1.50
		Modulus of elasticity in tension (GPa)		
Femur	25.5	25.0	14.9	17.6
Tibia	23.8	24.5	17.2	18.4
Humerus	17.8	18.3	14.6	17.5
Radius	22.8	25.9	15.8	18.9
		Ultimate compressive strength (MPa)		
Femur	145 ± 1.6	147 ± 1.1	100 ± 0.7	170 ± 4.3
Tibia	163	149 ± 1.4	106 ± 1.1	ND[b]
Humerus	154	144 ± 1.3	102 ± 1.6	ND
Radius	156	152 ± 1.5	107 ± 1.6	ND
		Ultimate percent contraction (%)		
Femur	2.4	1.7 ± 0.02	1.9 ± 0.02	1.85 ± 0.04
Tibia	2.2	1.8 ± 0.02	1.9 ± 0.02	ND
Humerus	2.0	1.8 ± 0.02	1.9 ± 0.02	ND
Radius	2.3	1.8 ± 0.02	1.9 ± 0.02	ND
		Modulus of elasticity in comoression (GPa)		
Femur	9.4 ± 0.47	8.7	4.9	ND
Tibia	8.5	ND	5.1	ND
Humerus	9.0	ND	5.0	ND
Radius	8.4	ND	5.3	ND
		Ultimate shear strength (MPa)		
Femur	99 ± 1.5	91 ± 1.6	65 ± 1.9	5.4 ± 0.6
Tibia	89 ± 2.7	95 ± 2.0	71 ± 2.8	ND
Humerus	90 ± 1.7	86 ± 1.1	59 ± 2.0	ND
Radius	94 ± 3.3	93 ± 1.8	64 ± 3.2	ND
		Torsional modulus of elasticity (GPa)		
Femur	16.3	16.8	13.5	3.2
Tibia	19.1	17.1	15.7	ND
Humerus	23.5	14.9	15.0	ND
Radius	15.8	14.3	8.4	ND

[a]Data are mean ±S.E.M. (Adapted from Yamada, 1970.)
[b]ND = no data.

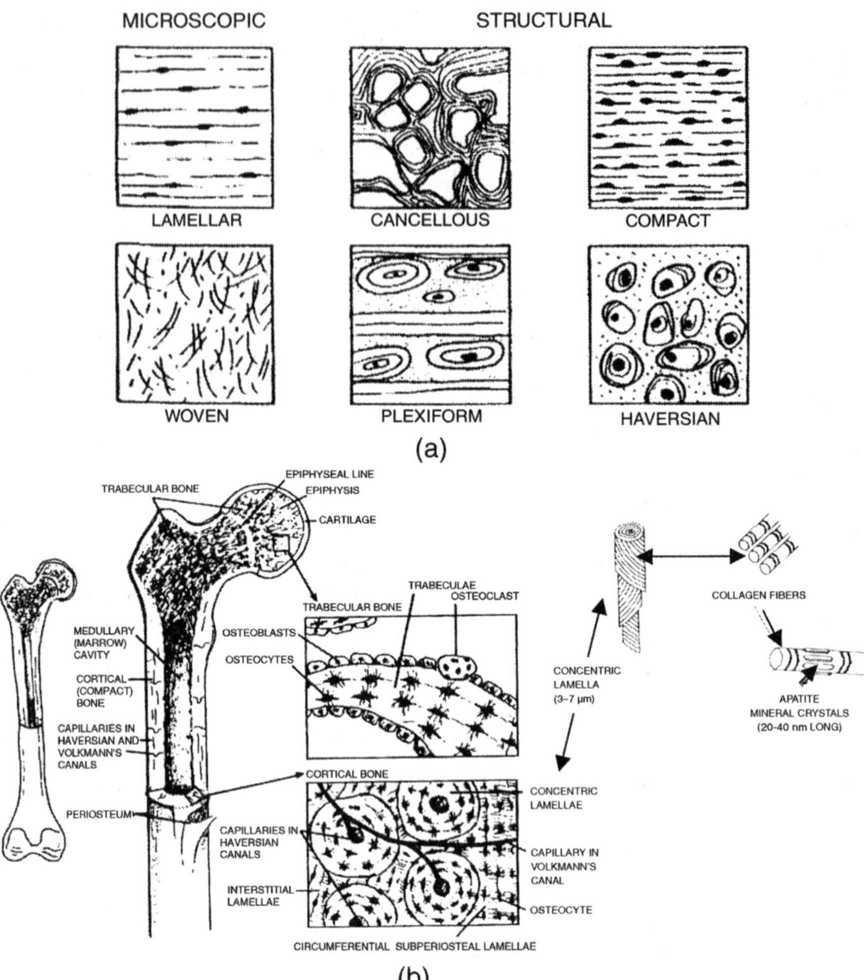

FIG. 1. Microarchitecture and structural classifications of physiological bone. (a) Schematic of microscopic and structural classifications of bone. (Redrawn and adapted from Kaplan *et al.,* 1994.) (b) Schematc of the microarchitecture of the femur. (Redrawn and adapted from Keaveny and Hayes, 1993, and Fung, 1993.)

contains no uniform orientation of collagen fibers; the isotropic mechanical characteristics (i.e., mechanical behavior independent of orientation of applied force) of woven bone are a consequence of its unoriented, nonuniform collagen fibers (Fung, 1993).

The other microarchitectural form of bone, lamellar bone, actively replaces maturing woven bone and, consequently, contains up to 100 times more mineralized matrix or hydroxyapatite; these hydroxyapatite crystals are 20 to 80 nm long and 2 to 5 nm thick in the human femur (Kaplan *et al.,* 1994; Park and Lakes, 1992). Lamellar bone is highly organized and contains stress-oriented collagen fibers; such orientation results in anisotropic mechanical properties (i.e., mechanical behavior dependent on the orientation of applied force) with greatest strength parallel to the longitudinal axis of the collagen fibers (Fung, 1993).

B. Structural Organization of the Bone Microarchitecture

Woven bone and lamellar bone are structurally organized into either trabecular (spongy or cancellous) or cortical (dense or compact) bone (Kaplan *et al.,* 1994; Park and Lakes, 1992). Trabecular bone, characterized by 50 and 90% porosity, contains large pores (up to several millimeters in diameter (Kaplan *et al,* 1994)). Trabecular bone, found primarily at the metaphyses and epiphyses of both long and cuboidal bones, is organized into a three-dimensional branching lattice with spicules of trabeculi oriented in the direction of principal stress; for trabecular bone, compression is the dominant force under physiological loading conditions (Fung, 1993).

Cortical bone, characterized by less than 30% porosity and composed of small pores (up to 1 mm in diameter), is classified as *haversian* bone and contains Volkmann's canals (canals in which capillaries reside) (Fig. 1b), lacunae, and canaliculi (Keaveny and Hayes, 1993; Fung, 1993). Cortical bone is found at the diaphyses of long bones and as circular envelopes in cuboidal bone (Keaveny and Hayes, 1993; Fung, 1993). In compact bone vascular channels are randomly oriented, in plexiform bone the vasculature is located in layers of woven bone dispersed within the lamellar bone, and in haversian bone (the most complex type of cortical bone) vascular channels circumferentially surround lamellar bone (Fig. 1a) (Kaplan *et al.,* 1994). The circular arrangements of bone around vascular channels found in haversian bone are called *osteons*. Osteons are usually oriented along the long axis of bone; the central canal of the osteon (called the *haversian canal*) contains hematopoietic cells, capillaries, and, occasionally, nerves (Keaveny and Hayes, 1993; Fung, 1993). The capillaries in the haversian canal are derived from the principal nutrient arteries of the epiphyseal and metaphyseal arteries of cortical bone and supply oxygen and nutrients necessary to maintain bone homeostasis (Park and Lakes, 1992).

C. Chemical Composition of the Bone Matrix

The human femur is a composite material; approximately 70% of its matrix is inorganic hydroxyapatite, 20% is organic, and 10% is water (Kaplan *et al.,* 1994; Fung, 1993). The detailed composition of bone depends on species, age, anatomical location, dietary history, and either the absence or presence of disease (Kaplan *et al.,* 1994; Park and Lakes, 1992).

1. Inorganic Phase

The inorganic component of bone is primarily platelike (20 to 80 nm long and 2 to 5 nm thick) crystalline hydroxyapatite, $Ca_5(PO_4)_3(OH)$ or HA (Kaplan *et al.,* 1994; Park and Lakes, 1992). Small amounts of impurities may be present in the mineralized HA matrix; for example, carbonate may replace phosphate groups, whereas chloride and fluoride may replace hydroxyl groups. Because release of ions from the mineral bone matrix controls cell-mediated functions, the presence of impurities may impact important biological aspects (and, subsequently, affect chemical and mechanical properties of bone) that are critical to normal bone function; for example, impurities present in the mineralized matrix may affect cellular function(s) that influence new bone formation (Kaplan *et al.,* 1994; Park and Lakes, 1992).

2. Organic Phase

Approximately 90% of the organic phase of bone is Type I collagen; the remaining 10% consists of noncollagenous proteins, such as lipids and other macromolecules (i.e., growth factors, cytokines, and adhesive proteins) (Kaplan *et al.,* 1994; Fung, 1993; Park and Lakes, 1992). Growth factors and cytokines (such as insulinlike growth factors and osteogenic proteins), proteins contained in serum, bone-inductive proteins (such as osteonectin, osteopontin, and osteocalcin), and extracellular matrix compounds (such as bone sialoprotein, bone proteoglycans, and other phosphoproteins as well as proteolipids) are present in the mineralized matrix and may mediate bone–cell function such as formation of new bone by osteoblasts and bone resorption by osteoclasts (Kaplan *et al.,* 1994).

a. *Collagen.* Distribution of the various types of collagen in the human body is tissue specific (Ayad *et al.,* 1994). Collagen (mainly Type I) found in bone is synthesized by osteoblasts (the bone-forming cells) and is secreted as triple helical procollagen into the extracellular matrix, where collagen molecules are stabilized by cross-linking of reactive aldehydes among the collagen chains. Each of the 12 types of collagen found in the body consists of 3 polypeptide chains composed of approximately 1400 amino acids. For example, Type I collagen (molecular weight 139,000) possesses 2 identical $\alpha 1(I)$ chains and 1 unique $\alpha 2$ chain; this configuration results in a

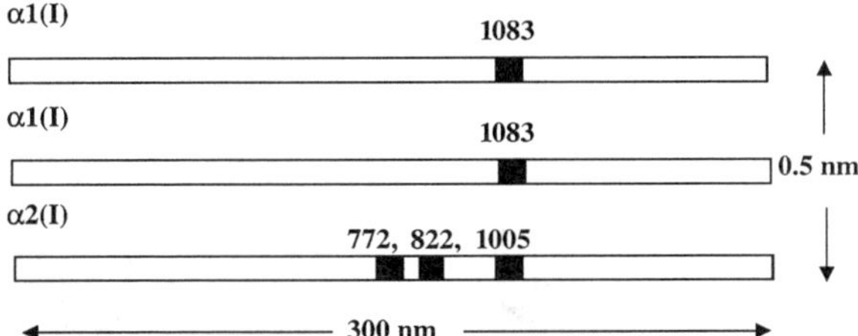

FIG. 2. Schematic representation (not to scale) of the amino acid sequence of a Type I collagen. A Type I collagen consists of two identical of $\alpha1$(I) and one unique $\alpha2$(I) chains whose combination produces a triple helical structure 300 nm long and 0.5 nm thick, with a periodicity of 67 nm. Arginine–glycine–aspartic acid adhesive peptides (■) start at peptide sequence 1083 in the $\alpha1$(I) chain as well as at 772, 822, and 1005 in the $\alpha2$(I) chain. (The schematic was redrawn using information obtained from the following references: Ayad *et al.*, 1994; Mathews and van Holde, 1990; and Darnell *et al.*, 1990.)

linear molecule that is 300 nm long (Fig. 2) (Ayad *et al.*, 1994; Mathews and van Holde, 1990). The linear molecules (or fibers) of Type I collagen are grouped in triple helical bundles having a periodicity of 67 nm, with gaps (called *hole-zones*) between the ends of the molecules and pores between sides of parallel molecules. During new bone formation, noncollagenous proteins and mineral are synthesized by osteoblasts and deposited into the hole-zones and pores of the collagen matrix. Type I collagen has cell-adhesive properties, particularly for osteoblasts (the bone-forming cells) (Steele *et al.*, 1993), due to the presence of the adhesive peptide arginine–glycine–aspartic acid (RGD) starting at amino acid sequence 1083 in the $\alpha1$(I) chain and at sequences 777, 822, and 1005 in the $\alpha2$(I) chain.

b. *Noncollagenous Proteins.* Numerous noncollagenous proteins are found in the bone matrix; osteocalcin, osteonectin, alkaline phosphatase, osteopontin, and bone sialoprotein (discussed in the section "Osteoblasts: The Bone-Forming Cells") are synthesized by osteoblasts. Other proteins, such as laminin, fibronectin, and vitronectin (discussed in the following sections) are found in blood serum as well as in the extracellular matrix of fibrous tissue and bone.

(i) *Laminin.* Laminin is a family of large glycoproteins that are distributed ubiquitously in the basement membrane of tissues; laminin has been shown to perform key roles in the development, differentiation, and migration of cells (primarily endothelial cells, cells that line the vasculature of the body, and neuronal cells, cells of the central nervous system) (Ayad *et al.*, 1994). There are four isomeric forms of laminin, specifically, A–S–B2, M–B1–B2,

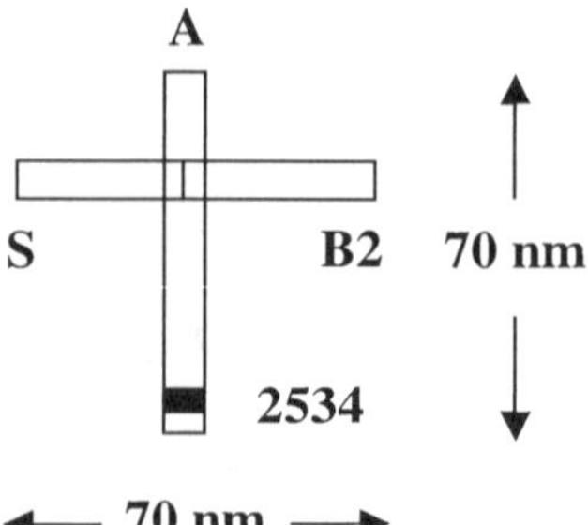

FIG. 3. Schematic representation (not to scale) of the amino acid sequence of the A–S–B2 isomer of laminin. Laminin is composed of one long arm (A) and two short arms (S and B2) arranged in a cruciform configuration 70 nm long and 70 nm wide. The arginine–glycine–aspartic acid adhesive peptide (■) starts at peptide sequence 2534 in chain A. (The schematic was redrawn using information obtained from the following references: Ayad *et al.,* 1994; Mathews and van Holde, 1990; and Darnell *et al.,* 1990.)

and M–S–B2; a truncated form of the B2, B2t chain was recently discovered (Ayad *et al.,* 1994). Distribution of these isomeric forms of laminin in the body is associated with developmental stages and is tissue specific. The laminin (molecular weight of 820,000) molecule has a cruciform configuration with one long arm (approximately 70 nm long) and two short arms (each approximately 35 nm long; Fig. 3) (Darnell *et al.,* 1990). Laminin has cell-adhesive properties, particularly for endothelial cells (Graf *et al.,* 1987), due to the presence of the following adhesive domains: peptides arginine–glycine–aspartic acid and isoleucine–lysine–valine–alanine–valine (IKVAV), which are located in chain A and start at peptide sequences 2534 and 2116, respectively; peptides leucine–glycine–threonine–isoleucine–proline–glycine (LGTIPG), arginine–tyrosine–valine–valine–leucine–proline–arginine (RYVVLPR), proline–aspartic acid–glycine–serine–glycine–arginine (PDGSGR), and tyrosine–isoleucine–serine–arginine (YIGSR), which are located in chain B1 and start at peptide sequences 463, 662, 923, and 950, respectively; and peptide leucine–arginine–glutamic acid (LRE), which is located in chain S and starts at sequence 1705 (Ayad *et al.,* 1994).

(ii) *Fibronectin.* Fibronectin (molecular weight 273,715) is a widely distributed glycoprotein present at high concentrations in bone matrix, plasma (for example, 300 μg/ml of human plasma), and in other body fluids and tissues (Ayad *et al.,* 1994). The principal functions of fibronectin are to mediate cell migration during development and wound healing, regulate cell growth and differentiation, and participate in haemostasis and thrombosis (Ayad *et al.,* 1994). Body-fluid fibronectin is a dimer of two identical subunits (each 60 to 70 nm long and 2 to 3 nm wide) covalently linked near the carboxyl termini by a pair of disulfide bonds (Fig. 4) (Darnell *et al.,* 1990).

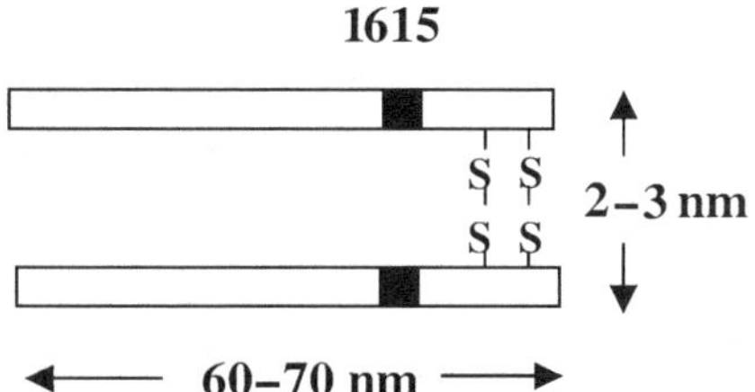

FIG. 4. Schematic representation (not to scale) of the amino acid sequence of fibronectin. Fibronectin is a dimer of two identical subunits (60 to 70 nm long and 2 to 3 nm long) covalently linked by a pair of disulfide bonds. The arginine–glycine–aspartic acid adhesive peptide (■) starts at peptide sequence 1615. (The schematic was redrawn using information obtained from the following references: Ayad *et al.,* 1994; Mathews and van Holde, 1990; and Darnell *et al.,* 1990.)

Fibronectin is an important adhesive protein (primarily for fibroblasts, cells that contribute to callus formation and fibrous encapsulation events that lead to orthopedic and dental implant loosening and eventual failure, but also for osteoblasts, the bone-forming cells (Thomas *et al.,* 1997) and various other cells) due to the presence of a number of bioactive domains such as arginine–glycine–aspartic acid, isoleucine–aspartic acid–alanine–proline–serine (IDAPS), leucine–aspartic acid–valine (LDV), arginine–glutamic acid–arginine valine (REDV), which start at peptide sequences 1615, 1994, 2102, and 2182, respectively (Ayad *et al.,* 1994).

(iii) *Vitronectin.* Vitronectin, found in both plasma and the extracellular matrix of bone, participates in a variety of physiological processes, including hemostasis, phagocytosis, tissue repair, and immune response. Vitronectin exists in two forms: a single chain (molecular weight of 75,000) and an endogenously clipped, two-chain form held together by disulfide bonds (molecular weights of approximately 65,000 and 10,000, respectively) (Ayad *et al.,* 1994). Structurally, vitronectin is asymmetrically shaped (total length approximately 15 nm; Fig. 5) with a large content of β-pleated sheets (Ayad *et al.,* 1994). Vitronectin plays an important role in adhesion and migration

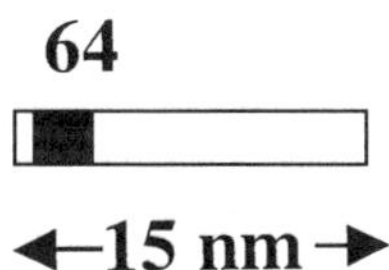

FIG. 5. Schematic representation (not to scale) of the amino acid sequence of vitronectin. Vitronectin is 15 nm long and possesses the arginine–glycine–aspartic acid adhesive peptide (■) starting at peptide sequence 64. (The schematic was redrawn using information obtained from the following references: Ayad *et al.,* 1994; Mathews and van Holde, 1990; and Darnell *et al.,* 1990.)

of various cells, particularly osteoblasts, the bone-forming cells (Steele *et al.*, 1993; Healy *et al.*, 1994). The vitronectin molecule contains the arginine–glycine–aspartic acid peptide starting at sequence 64 (Ayad *et al.*, 1994).

D. Cells of the Bone Tissue

1. Osteoblasts: The Bone-Forming Cells

It is well established that osteoblasts and, to a lesser extent, osteocytes (mature osteoblasts) contribute to new bone synthesis (Keaveny and Hayes, 1993; Kaplan *et al.*, 1994). The principal difference between osteoblasts and osteocytes is their relative location in bone. Osteoblasts are located on the periosteal and endosteal surfaces of bone (Fig. 1(b)). Once an osteoblast becomes surrounded by a mineralized matrix, however, the cell is characterized by a higher nucleus–to–cell cytoplasm ratio and becomes known as an osteocyte (Kaplan *et al.*, 1994). Osteocytes are arranged concentrically around the central lumen of an osteon and in between lamellae (Fig. 1(b)). Osteocytes possess extensive cell processes with which they establish contacts with adjacent osteocytes through small channels, or *canaliculi,* present in bone. Due to their three-dimensional distribution and interconnecting cell processes, osteocytes are believed to play a pivotal role in communicating physiological stress and strain signals in bone tissue (Kaplan *et al.*, 1994). Osteocytes regulate new bone formation by modulating osteoblast function through secretion of growth factors such as the insulinlike growth factor I and the tissue growth factor β (Kaplan *et al.*, 1994; Trippel, 1998). These growth factors mediate the differentiation of osteoblasts from immature, non-calcium-depositing cells to mature osteoblasts that deposit calcium-containing mineral into the extracellular matrix of bone. Phenotypic markers of the differentiation of non-calcium-depositing osteoblasts to osteoblasts that deposit calcium-containing mineral *in vitro* have been well studied (Stein and Lian, 1993; Stein *et al.*, 1990); these studies have provided much evidence concerning *in vivo* functions of osteoblasts leading to the synthesis and deposition of bone on newly implanted prostheses.

Three distinct periods of osteoblast differentiation at the genetic level have been identified during *in vitro* examination of developing osteoblasts after initial adhesion to a surface: (1) cell proliferation and extracellular matrix synthesis, (2) extracellular matrix development and maturation, and (3) extracellular matrix mineralization (Stein and Lian, 1993). A schematic of the time course of osteoblast function and synthesis of extracellular matrix proteins on a newly implanted biomaterial is shown in Fig. 6.

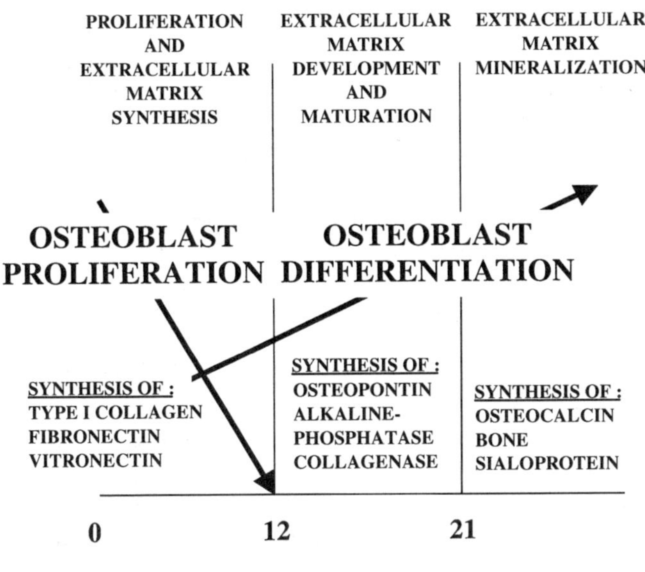

FIG. 6. Time course of osteoblast function and synthesis of extracellular matrix proteins on a newly implanted biomaterial. After initial adhesion of osteoblasts to a newly implanted biomaterial, three distinct phases of osteoblast differentation occur : (1) proliferation and extracellular matrix synthesis, (2) extracellular matrix development and maturation, and (3) extracellular matrix mineralization. (Adapted and redrawn from Stein and Lian, 1993.)

After initial adhesion to a biomaterial device, osteoblasts actively proliferate and express genes for Type I collagen, fibronectin, and vitronectin (Stein *et al.*, 1990). Osteoblasts also express genes for osteopontin, an acidic glycoprotein that possesses several calcium-binding sites and mediates adhesive interactions of these cells with the extracellular matrix (Stein and Lian, 1993). Genes for osteopontin synthesis are expressed minimally during the proliferative stages but in higher quantities late in the extracellular matrix development and maturation stages of osteoblast development (Stein *et al.*, 1990).

As the proliferative phase ends and the extracellular matrix development and maturation phase begins, alkaline phosphatase activity (expressed by numerous cells, including osteoblasts, fibroblasts, and leukocytes as well as by bone marrow reticular cells (Gehron-Robey, 1989)) and mRNA expression for proteins (such as osteocalcin, bone sialoprotein, and collagenase) associated with the osteoblastic phenotype are increased tenfold (Stein and Lian, 1993); for example, expression of the collagenase (an enzyme that aids in collagen turnover during reorganization and maturation of the extracellular matrix) gene is upregulated as the proliferative phase ends and the extracellular matrix mineralization phase commences (Stein and Lian, 1993).

Bone sialoprotein, osteopontin, and osteocalcin are synthesized and deposited as the mineralization process begins and mineral nodules form (Stein and Lian, 1993). Bone sialoprotein contains the cell-adhesive arginine–glycine–aspartic acid peptide sequence and may thus mediate osteoblast adhesion on the extracellular matrix (Gehron-Robey, 1989). Osteocalcin, a calcium-binding protein, interacts with hydroxyapatite and is thought to mediate coupling of bone resorption (by osteoclasts) and bone formation (by osteoblasts and/or osteocytes) (Stein and Lian, 1993).

2. *Osteoclasts: The Bone-Resorbing Cells*

Osteoclasts, cells primarily responsible for resorption of bone, are distinguished by their large (20 to 100 μm in diameter), multinuclear morphology (Fig. 1(b)). Osteoclasts are derived from pluripotent (i.e., capable of differentiating into various cells, including monocytes and macrophages) cells of the bone marrow (Kaplan *et al.,* 1994; Park and Lakes, 1992). Osteoclasts resorb bone by forming ruffled cell membrane edges (thereby increasing their surface area of attachment onto bone surfaces), lowering the pH of the local environment by producing hydrogen ions through the carbonic anhydrase system (thus increasing the solubility of hydroxyapatite crystals, the major inorganic component of bone), and, lastly, by removing organic components of the matrix via acidic proteolytic digestion that results in the formation of bone resorption pits termed *Howship's lacunae* (Kaplan *et al.,* 1994; Park and Lakes, 1992). The bone-resorbing activity of osteoclasts is believed to instigate the formation of new bone by osteoblasts, as discussed in the section "Bone Remodeling" (Rifkin and Gay, 1992; Stein and Lian, 1993; Heegard, 1993).

3. *Fibroblasts*

Fibroblasts are derived from mesenchymal cells of the bone marrow (Kaplan *et al.,* 1994; Park and Lakes, 1992), are found unbiquitously in various tissues (such as skin, vasculature, lungs, and bone), and are primilarily responsible for the formation of fibrous, connective tissue (Darnell *et al.,* 1990). Fibroblasts participate in maintaining the mechanical integrity of bone by synthesizing and secreting collagen into the extracellular matrix (Kaplan *et al.,* 1994). However, excessive secretion of fibrous tissue and/or callus formation during new bone formation may compromise the mechanical properties (such as bending) of bone. Moreover, fibrous encapsulation and callus formation around newly implanted orthopedic or dental prostheses are the most frequent causes of incomplete osseointegration between an implant and juxtposed bone (Brunski, 1991); incomplete osseointegration may lead to implant loosening and eventual failure. For these reasons, functions (such as adhesion, synthesis and secretion of fibrous tissue) of fibroblasts should be minimized at the surface of a newly implanted orthopedic or dental device.

E. Bone Remodeling

Both cortical and trabecular bone are continuously remodeled through the formation of a bone-modeling unit (BMU), or *cutter-cone;* this process involves activation of osteoclasts, leading to resorption of bone by osteoclasts and formation of new bone by osteoblasts on the site of the "old," resorbed bone (Fig. 7) (Martin and Burr, 1989). Under normal physiological conditions (i.e., in the absence of either growth or disease) the dynamics of bone remodeling maintain bone homeostasis throughout a person's lifetime.

Nutrients necessary for bone remodeling, mesenchymal stem cells (pluripotent cells capable of differentiating into various cells such as osteoblasts, endothelial cells, and fibroblasts), and hematopoeitic cells (including cells

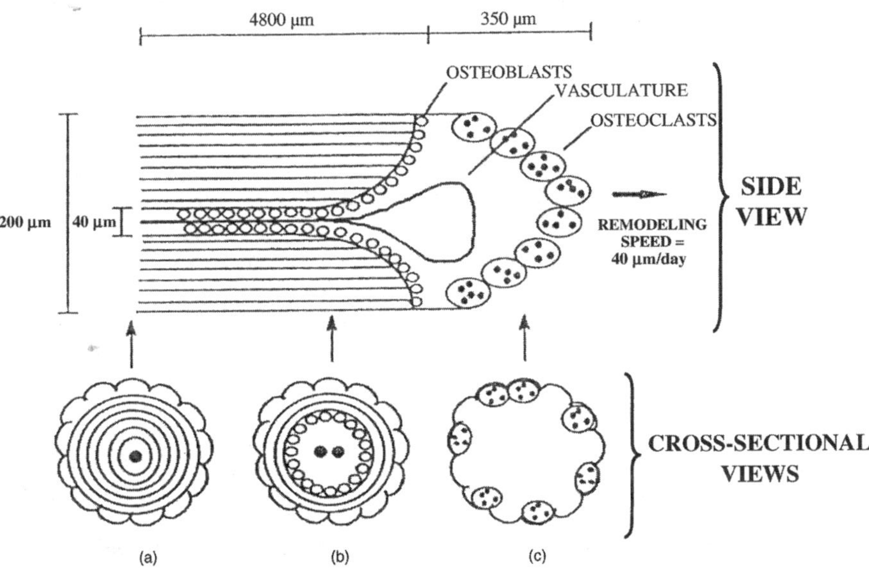

FIG. 7. Schematic diagram of a bone modeling unit, or cutter-cone. Cross-sectional views ((a), (b), and (c)) match respective side view sections of a bone modeling unit. (Adapted and redrawn from Martin and Burr, 1989.)

from the monocyte or macrophage cell line capable of differentiating into osteoclasts) are supplied to bone through the vascular network of the bone marrow. Osteoclasts (either on or in the bone as well as those supplied by the vascular network) are activated by growth factors, cytokines, and proteins present in the bone matrix to resorb old bone (Martin and Burr, 1989). Osteoblasts are then activated by growth factors (such as insulin-like growth factors I and II) secreted by osteoclasts to deposit calcium-containing mineral (Kaplan *et al.,* 1994). Mineral accretion in skeletal tissue arises by (1) nucleation or initial recruitment and deposition of precipitating ions (such as calcium, phosphate, and other ions) into pores and/or *hole zones* of the extracellular collagen matrix that had been secreted by osteoblasts, (2) hydroxyapatite (HA) crystal growth promoted by noncollagenous proteins (such as phosphoproteins, osteonectin–collagen complexes, and proteolipids), and (3) secondary nucleation, in which new crystals of hydroxyapatite are deposited on nuclei of existing hydroxyapatite (Kaplan *et al.,* 1994; Park and Lakes, 1992). Extracellular matrix proteins, growth factors, and cytokines are also believed to control the size (20 to 80 nm in diameter) as well as extent of HA crystal growth by preventing further deposition of mineral by osteoblasts according to mechanisms that have yet to be fully understood (Kaplan *et al.,* 1994; Keaveny and Hayes, 1993).

The extent of bone remodeling is influenced by a number of factors, including but not limited to dietary history, exercise frequency, age, injury, and/or the presence of a prosthetic, manufactured device (Kaplan *et al.,* 1994; Park and Lakes, 1992). More importantly, the extent of bone remodeling that occurs at an implant surface will determine the fate of the prosthetic device; for example, loosening and failure of the orthopedic or dental implant may result from either (1) little or no remodeling in the bone surrounding an implant, which may lead to malnourished juxtaposed bone, or (2) enhanced remodeling in the bone surrounding an implant, which may lead to excessive bone resorption, or osteolysis (Brunski, 1991). Events that occur at the tissue–implant interface will, clearly, control the extent of bone remodeling around the prostheses and, therefore, integration of the biomaterial into surrounding bone.

IV. The Tissue–Implant Interface

Implantation of a synthetic material into mammalian living tissue causes a number of biological host responses, including rejection by the body, encapsulation in newly formed fibrous tissue, and successful incorporation into surrounding tissues. The fate of an implanted device is determined by cellular or molecular events at tissue–implant interfaces; these events are mediated by the wound-healing process of bone in response to surface properties of the prostheses.

A. Wound-Healing Response of Bone

Implanting or introducing a biomaterial into the body by surgical procedures inevitably causes damage to surrounding tissues, and, consequently, initiates a series of host responses. Inflammation and the wound-healing process involve recruitment of a variety of cell types, body fluids, and proteins to the tissue–implant interface (Anderson, 1993; Hench and Ethridge, 1975). For orthopedic and dental implants, select osteoblast (bone-forming cells) recruitment to the implant-material surface is imperative for subsequent formation of new bone, leading to successful osseointegration. Bonding of orthopedic and dental implants to surrounding bone provides mechanical stability to the prostheses *in situ* and minimizes motion-induced trauma to surrounding tissues; formation of callus (instead of bony) tissue at these implantation sites decreases implant efficacy and may eventually result in clinical failure (Brunski, 1991).

B. Protein Interactions with Biomaterial Surfaces

Specific domains of proteins (for example, those mentioned in the section "Organic Phase") adsorbed to biomaterial surfaces interact with select cell membrane receptors (Fig. 8); accessibility of adhesive domains (such as specific amino acid sequences) of select adsorbed proteins may either enhance or inhibit subsequent cell (such as osteoblast) attachment (Schakenraad, 1996). Several studies have provided evidence that properties (such as chemistry, charge, and topography) of biomaterial surfaces dictate select interactions (such as type, concentration, and conformation or bioactivity) of plasma proteins (Sinha and Tuan, 1996; Horbett, 1993; Horbett, 1996; Brunette, 1988; Davies, 1988; Luck *et al.*, 1998; Curtis and Wilkinson, 1997). Albumin has been the protein of choice in protein-adsorption investigations because of availability, low cost (compared to other proteins contained in serum), and, most importantly, well-documented conformation or bioactive structure (Horbett, 1993); recently, however, a number of research groups have started to examine protein (such as fibronectin and vitronectin) interactions with material surfaces that are more pertinent to subsequent cell adhesion (Luck *et al.*, 1998; Degasne *et al.*, 1999; Dalton *et al.*, 1995; Lopes *et al.*, 1999).

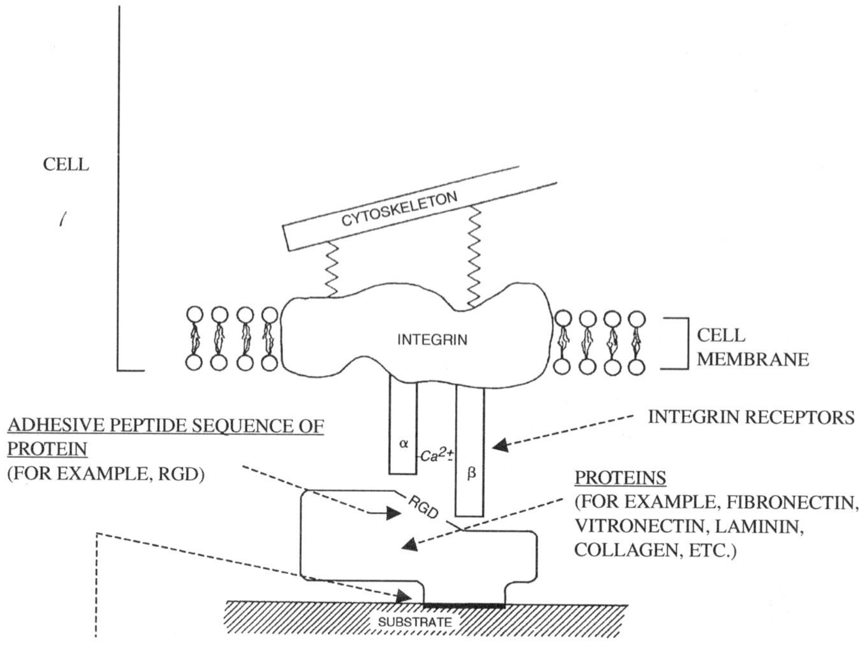

FIG. 8. Schematic representation of protein-mediated cell adhesion on biomaterial surfaces. Biomaterial surface properties (such as hydrophilicity/hydrophobicity, topography, energy, and charge) affect subsequent interactions of adsorbed proteins; these interactions include but are not limited to adsorbed protein type, concentration, and conformation. Changes in protein–surface interactions may alter accessibility of adhesive domains (such as the peptide sequence arginine–glycine–aspartic acid) to cells (such as osteoblasts, fibroblasts, or endothelial cells) and thus modulate cellular adhesion. (Adapted and redrawn from Schakenraad, 1996.)

It has been reported in the literature that changes in the type and concentration of protein [specifically, albumin (Luck *et al.,* 1998), fibronectin (Degasne *et al.,* 1999), and vitronectin (Dalton *et al.,* 1995; Lopes *et al.,* 1999)] adsorption on material surfaces depends on a material's surface properties, such as chemistry (i.e., either polymer, metal, or ceramic), hydrophilicity and hydrophobicity, roughness, and surface energy. Specifically, maximum vitronectin (Lopes *et al.,* 1999), fibronectin (Degasne *et al.,* 1999) and albumin (Horbett, 1993; Luck *et al.,* 1998) adsorption was noted on hydrophilic surfaces with high surface roughness and/or energies.

Recent studies have attempted to further elucidate mechanisms of protein adsorption on biomaterial surfaces. For example, Ellingsen (1991) reported that adsorption of calcium on titanium surfaces subsequently enhanced binding of select proteins. In contrast, adsorption of other ions (such as magnesium) contained in physiological fluids on titanium surfaces did not affect subsequent select protein adsorption (Ellingsen, 1991).

C. Protein-Mediated Cell Adhesion on Biomaterial Surfaces

Select proteins that mediate adhesion of specific anchorage-dependent cells (such as osteoblasts, fibroblasts, and endothelial cells) on substrate surfaces have been identified (Underwood and Bennett, 1989; Thomas *et al.,* 1997; Ayad *et al.,* 1994). For example, adsorption of fibronectin and vitronectin on tissue-culture polystryene subsequently enhanced osteoblast, fibroblast, and endothelial cell adhesion (Underwood and Bennett, 1989). More importantly, fibronectin and vitronectin adsorption on borosilicate glass, in a competitive environment, maximized fibroblast and osteoblast adhesion, respectively (Thomas *et al.,* 1997). Ayad *et al.* (1994) reported that enhanced adsorption of laminin on tissue-culture polystyrene promoted subsequent endothelial cell adhesion. These studies provided evidence that adsorption of specific protein(s) can, subsequently, control select cell adhesion on material surfaces.

Cells interact with their external environment through signals (such as chemical, electrical, and mechanical) transmitted through the cell membrane. For this reason, understanding cellular interactions with a biomaterial surface requires elucidation of molecular processes that occur at the cell membrane–biomaterial interface. For example, cellular adhesion (a crucial prerequiste function for anchorage-dependent cells such as osteoblasts) has been well examined at the molecular level. Cell-binding regions of extracellular matrix proteins (such as the arginine–glycine–aspartic acid peptide sequence present in vitronectin, fibronectin, collagen and laminin, as discussed in the section "Organic Phase") and respective cell-membrane-intercalated receptors (i.e., the integrins) have been identified as being among the most important mechanisms for cell adhesion to substrates, including borosilicate glass and tissue-culture polystyrene (Kramer *et al.,* 1993).

Integrins are a family of transmembrane heterodimeric glycoproteins that are receptors for specific epitopes of extracellular matrix proteins and for other cell-surface molecules (Kramer *et al.,* 1993). Integrins exist as a dimer complex composed of an α-subunit (120–180 kD) noncovalently associated with a β-subunit (90–110 kD) (Hynes, 1992). At least 8 β-subunits and 14 α-units have been identified and are concentrated at loci, called focal adhesion sites, of close proximity between cells and extracellular matrices on substrates (Hynes, 1992). Focal adhesion sites are points of aggregation of, and are physically associated with, intracellular cytoskeletal molecules that control, direct, and modulate cell function in response to extracellular signals (Schwartz, 1992).

A number of integrins bind to the RGD sequence of extracellular matrix proteins; for example, integrins $\alpha_v\beta_1$ and $\alpha_v\beta_3$ bind to the RGD epitope (ligand) of vitronectin, fibronectin, and Type I collagen (Cox *et al.,* 1994). Expression of integrins on human osteoblasts is mediated by surface properties (such as topography and composition) of substrates on which these cells adhere (Gronowicz and McCarthy, 1996; Schneider and Burridge, 1994; Sinha and Tuan, 1996); specifically, expression of the $\alpha_v\beta_5$ and $\alpha_v\beta_6$ integrins was enhanced when the cells adhered on surfaces of increased roughness (e.g., sandblasted titanium substrates) and on smooth surfaces (e.g., grit-polished titanium substrates), respectively (Sinha and Tuan, 1996). Because the $\alpha_v\beta_5$ integrin has been exclusively associated with binding to the RGD protein ligand of vitronectin, it can be suggested that vitronectin plays a critical role in the adhesion of osteoblasts on rough surfaces. Furthermore, $\alpha_v\beta_6$ has been exclusively associated with binding to the RGD protein ligand of fibronectin; for this reason, it can also be suggested that fibronectin is an important protein for osteoblast adhesion on smooth surfaces.

In addition to mediating cell adhesion, it has been demonstrated that integrin expression by osteoblasts determines their phenotypic expression (see the section "Osteoblasts: The Bone-Forming Cells"). For example, addition of either soluble concentrations of RGD (Moursi *et al.,* 1996) or antibodies of the integrin pair $\alpha_v\beta_6$ (Moursi *et al.,* 1997) to confluent osteoblast cultures, blocked initiation and formation of mineral nodules; these results provided evidence that integrin–fibronectin interactions leading to extracellular matrix development play a crucial role in osteoblast function (Moursi *et al.,* 1996, 1997).

However, integrin–RGD interactions are not the only mechanisms by which osteoblasts adhere. Several articles suggested that *in vivo* (Nakamura and Ozawa, 1994) and *in vitro* (Puleo and Bizios, 1992; Dalton *et al.,* 1995) osteoblasts attach to an implanted material through cell membrane heparan sulfate proteoglycan interactions with, for example, heparin-binding sites on fibronectin and collagen. Nakamura and Ozawa (1994) immunohistochemically detected heparan sulfate on the membranes of osteoblasts attached to bone matrix. In addition, by blocking heparin-binding sites of fibronectin (with platelet factor IV), osteoblast adhesion on fibronectin was inhibited by 45% (Puleo and Bizios, 1992). In this manner, in addition to integrin–RGD interactions, it has been demonstrated that osteoblast adhesion is also mediated by cell membrane heparan sulfate (Laterra *et al.,* 1983; Izzard *et al.,* 1986).

Reports found in the literature suggested that the peptide sequence lysine–arginine–serine–arginine (KRSR) selectively enhanced osteoblast adhesion by possibly binding to heparan sulfate on the membranes of osteoblasts (Dee *et al.,* 1996). Compared to unmodified glass, Dee (1996) demonstrated enhanced osteoblast, fibroblast, and endothelial cell adhesion on borosilicate glass modified with the immobilized integrin-binding RGD peptide. In contrast, compared to unmodified borosilicate glass, osteoblast adhesion was significantly greater, whereas fibroblast and endothelial cell adhesion were similar on borosilicate glass modified with the immobilized heparan–sulfate-binding KRSR peptide (Dee, 1996; Dee *et al.,* 1996). For this reason, Dee *et al.* (1996) suggested that proactive orthopedic and dental implants must enhance osteoblast adhesion by both integrin and heparan sulfate mechanisms; this could be accomplished by modifying the surface of a biomaterial with both integrin-binding peptides containing the RGD sequence and heparan sulfate-binding peptides containing the KRSR sequence (Dee, 1996; Dee *et al.,* 1996).

V. Materials Currently Used as Orthopedic and Dental Implants

Metals (such as commercially pure titanium, titanium alloys, stainless steel, and Co–Cr alloys), ceramics (such as alumina, hydroxyapatite, and bioglass), and polymers (such as polymethyl methacrylate (PMMA) and ultrahigh-molecular-weight polyethyene) have been used as either single or multicomponent orthopedic and dental implants (Park and Lakes, 1992). Traditionally, materials (such as metals, ceramics, and polymers) currently used for orthopedic and dental implant applications were selected by trial-and-error processes; to date, titanium and titanium alloys have been the orthopedic and dental material of choice due to their superior mechanical properties (such as low-weight–high-strength ratio), excellent resistance to corrosion, and ability to remain inert *in vivo* (Lausmaa *et al.,* 1990). Although satisfying (but not matching; see Table II) mechanical requirements of bone, titanium and titanium alloys have failed clinically, often due to insufficient bonding to juxtaposed bone—that is, incomplete osseointegration (Kaplan *et al.,* 1994).

Bioceramics, such as alumina or aluminum oxide (Al_2O_3), hydroxyapatite ($Ca_5(PO_4)_3(OH)$), calcium phosphates, bioglass–ceramics, and carbon-based alloys, possess exceptional biocompatability with bone cells and tissues; mechanical properties (such as low ductility), however, have limited their wide use as orthopedic and dental implant materials (Table II) (Doremus, 1992; Grenoble *et al.,* 1972). Once implanted into bone (usually as coatings for traditional orthopedic or dental implant metals), various bioceramics (such as hydroxyapatite, bioactive glass, and calcium phosphate) physiochemically bonded to bone and, in select cases, promoted new bone formation, leading to implant osseointegration (Jarcho *et al.,* 1977; Garcia and Doremus, 1992; Hench, 1993; Ducheyne, 1994, 1999). Direct chemical bonding of hydroxyapatite-coated titanium implants to adjacent bone

TABLE II
COMPARISON OF MECHANICAL PROPERTIES OF SELECT ORTHOPEDIC AND DENTAL MATERIALS AND BONE

Material	Modulus of elasticity (GPa)
Bone[a]	
Cortical bone	17
Trabecular bone	100
Metals[b]	
Titanium	110
Steel alloy	21
Aluminum alloy	7
Ceramics[c]	
Calcium phosphate	40–117
Alumina	380
Polymers[b]	
Polymethyl methacrylate (PMMA)	2

Data obtained from:
[a] (Fung, 1993).
[b] (Kaplan *et al.,* 1994).
[c] (Park and Lakes, 1992).

has been reported (Garcia and Doremus, 1992); the coordination between negatively charged carboxalate groups in collagen present in bone and calcium ions present on apatite surfaces has been proposed as the most likely ionic-bonding mechanism (Garcia and Doremus, 1992). Surface properties (such as chemical composition, specific surface area, crystal structure, and porosity) of bioceramics determine the rate, formation, and chemical composition of apatites synthesized on these material surfaces (Jarcho *et al.,* 1977; Garcia and Doremus, 1992; Hench, 1993; Radin and Ducheyne, 1993; Ducheyne, 1994; Pereira *et al.,* 1994; Yubao *et al.,* 1994). Deposition of calcium-deficient, carbonate-containing, bonelike apatite on the surface of bioceramics is thought to be the critical step in their osseointegration. Several studies have reported formation of apatite layers *in vitro* in the presence of acellular conditions that simulated the physiological milieu (that is, buffered saline solutions containing protein and ion concentrations similar to that of human plasma) (Radin and Ducheyne, 1993; Pereira *et al.,* 1994; Li *et al.,* 1994). In contrast to these studies, numerous reports in the literature demonstrated insufficient formation of bone on orthopedic and dental implant materials consisting of ceramics (such as calcium phosphates and hydroxyapatite) (Doremus, 1992). To date, the source of inconsistency in the efficacy of bioceramics remains largely unclear.

A. Novel Surface Modifications of Conventional Orthopedic and Dental Implants

Several attempts have been made to increase the cytocompatibility of conventional materials for orthopedic and dental applications using *in vitro* cellular models. Anchorage-dependent cells (such as osteoblasts, fibroblasts and endothelial cells) must first adhere to a surface in order to perform normal, subsequent functions (such as spreading and proliferation). Although adhesion is a critical cell process, subsequent functions (such as the deposition of calcium-containing mineral) by osteoblasts on orthopedic and dental implants are of tantamount importance. *In vitro* functions of osteoblasts have been well documented on conventional orthopedic and dental materials composed of metals (such as commercially pure titanium (Degasne *et al.,* 1999; Keller *et al.,* 1994; Wen *et al.,* 1996)), metal alloys (for example, Ti–6Al–4V (Puleo *et al.,* 1993), Co–Cr–Mo (Garvey and Bizios, 1995), and 316L stainless steel (Garvey and Bizios, 1995)) and ceramics (such as hydroxyapatite (Malik *et al.,* 1992) and bioglasses (Davies and Matsuda, 1994)). Reports in the literature demonstrated increased osteoblast adhesion, cell spreading, proliferation, synthesis of extracellular matrix proteins (such as alkaline phosphatase), and deposition of extracellular matrix calcium on conventional titanium surfaces with increased microsize roughness (i.e., sandblasted, heated treated, acid-etched, and machined surfaces) (Degasne *et al.,* 1999; Keller *et al.,* 1994; Wen *et al.,* 1996; Curtis and Wilkinson, 1997; Brunette, 1988). Besides promoting functions of osteoblasts, increased microsurface roughness of conventional ceramic surfaces (such as acid-etched hydroxyapatite) enhanced synthesis of tartrate–resistant–acid phosphatase (TRAP) and bone-resorption activity of osteoclast cells (Gomi *et al.,* 1993; Matsunaga *et al.,* 1999).

In addition to increased surface roughness, chemical modifications of materials currently used for orthopedic and dental implants have enhanced *in vitro* osteoblast functions such as adhesion, proliferation, synthesis of extracllular matrix proteins, and deposition of calcium-containing mineral (Healy *et al.,* 1994; Dee, 1996; Dee *et al.,* 1996). By immobilizing bioactive chemical compounds on functional groups of a silane-coated orthopedic or dental biomaterial, a traditional biomaterial may be transformed into one that can elicit specific responses from surrounding living cells and tissues. To date, enzymes (i.e., glucose oxidases, or glutamate oxidase (Tiller *et al.,* 1999)), antibodies (such as antimouse immunoglobulin G (Turkova, 1999)), and specific peptide sequences (such as arginine–glycine–aspartic acid–serine and lysine–arginine–serine–arginine (Dee, 1996; Dee *et al.,* 1996)) have been immobilized on various materials (i.e., glass, polymers, and metal oxides) to enhance osteoblast functions. It is not clear whether, upon implantation, the bioactivity of these immobilized peptides may be affected because of the interactions of macromolecules from physiological fluids and tissues; osseointegration of juxtaposed bone and biomaterial surfaces modified with immobilized specific bioactive groups remains to be proven *in vivo*. Clearly, new material formulations must be designed that retain their bioactivity *in vivo* to promote osseointegration with surrounding bone.

VI. Next Generation of Orthopedic and Dental Implants: Nanophase Ceramics

Traditional materials for orthopedic and dental applications have been selected based on their mechanical properties and ability to remain inert *in vivo;* this selection process has provided materials that satisfied physiological loading conditions but did not duplicate the mechanical, chemical, and architectural properties of bone. Most importantly, to date, failure of conventional orthopedic and dental implant materials is often due to insufficient bonding to juxtaposed bone (that is, insufficient osseointegration).

Biomaterial scientists and engineers are currently investigating novel formulations and modifications of existing materials that elicit specific, timely, and desirable responses from surrounding cells and tissues to support the osseointegration of the next generation of orthopedic and dental biomaterials (Ratner, 1992). Enhanced deposition of mineralized matrix at the bone–implant interface provides crucial mechanical stability to implants. Proactive orthopedic and dental biomaterials could consist of novel formulations that selectively enhance osteoblast function (such as adhesion, proliferation and formation of calcium-containing mineral) while, at the same time, minimize other cell (such as fibroblast) functions that may decrease implant efficacy (e.g., fibroblast participation in callus formation and fibrous encapsulation of implants *in vivo*).

Nanophase materials are new formulations of materials that are composed of grains of the same atoms but with fewer (less than tens of thousands) and smaller (less than 100 nm in diameter) atoms than in conventional forms (which contain several billion atoms and have grain sizes of micrometers to millimeters in diameter) (Siegel, 1996). Nanocrystalline materials exhibit enhanced magnetic, catalytic, electrical and optical properties when compared to conventional formulations of the same material (Siegel and Fougere, 1994, 1995a,b; Siegel, 1996, 1994). Moreover, nanophase ceramics can be synthesized so they possess similar grain size, geometry, and microarchitecture as that of healthy, physiological bone (see the sections "Microarchitecture" and "Structural Organization of the Bone Microarchitecture"). Properties (specifically, mechanical and surface) of nanophase materials pertaining to the design and synthesis of orthopedic and dental implants of increased efficacy are expanded upon in the sections that follow.

A. Surface Properties of Nanophase Ceramics for Enhanced Orthopedic and Dental Implant Efficacy

1. Rationale

Orthopedic and dental implants with surface properties that promote cell and tissue interactions that lead to implant osseointegration are needed. Surface properties (such as area, charge, and topography) depend on the grain size of a material; in this respect, nanophase materials, which, by their very nature, possess higher surface area with increased portions of surface defects (such as edge–corner sites) and grain boundaries (Klabunde *et al.,* 1996), have an advantage that currently remains largely unexplored for biomedical applications. The increased surface reactivity of nanomaterials has been utilized for catalytic applications (Klabunde *et al.,* 1996); for example, compared to conventional (greater than a 100-nm average grain size) magnesium oxide (MgO), nanophase (i.e., a 4-nm average grain size) materials, such as magnesium oxide (MgO) and aluminum nitride, possessed higher surface area (100 to 160 m^2/g compared to 200 to 500 m^2/g for MgO, respectively), less acidic OH^- groups (due to a much higher proportion of edge sites for the nanophase MgO to cause delocalization of electrons; Fig. 9), increased adsorption of acidic species (such as SO_2^- and CO_2^-), and increased destructive adsorption of organophosphorous and of chlorocarbons (Klabunde *et al.,* 1996; Baraton *et al.,* 1997). It is extremely attractive to ponder if and

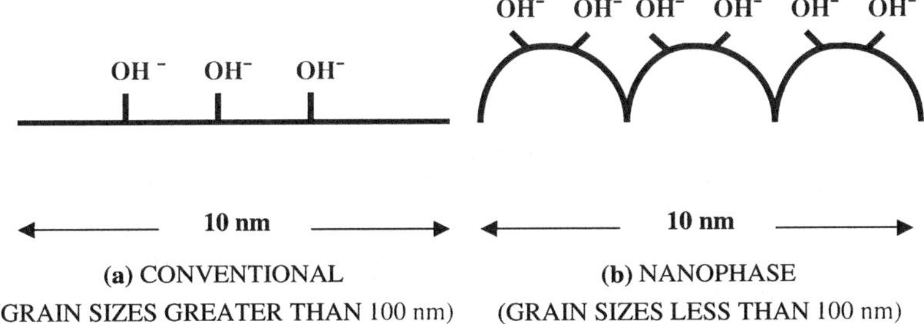

(a) CONVENTIONAL (GRAIN SIZES GREATER THAN 100 nm) (b) NANOPHASE (GRAIN SIZES LESS THAN 100 nm)

FIG. 9. Theoretical predictions of changes in surface properties of nanophase materials that affect the hydroxide layer. Compared to (a) conventional materials, (b) nanophase materials possess higher surface area and less acidic OH^- groups (due to an increase in electron delocalization) in the hydroxide layer. (Adapted and redrawn from Klabunde *et al.,* 1996.)

how these enhanced surface properties (such as increased surface area and charge, as well as ability to alter adsorption of chemical species) could be used to promote bonding of juxtaposed bone to an orthopedic or dental implant composed of nanophase ceramics.

2. *Experimental Evidence*

Clearly, some of the best examples illustrating the potential advantages of nanoceramic surface properties in advancing orthopedic and dental implant applications are calcium phosphate and/or calcium phosphate derivatives (such as hydroxyapatite ceramics, tricalcium phosphates, calcium carbonates, and bioactive glasses). As the main inorganic component of bone, hydroxyapatite (with a physiological platelike 25×3.5-nm geometry (Muller-Mai *et al.*, 1995)) has been an attractive material for hard-tissue repair over the last three decades. When implanted, these materials (sometimes polygonal coarse particles less than 100 nm in diameter (Muller-Mai *et al.*, 1995)) are nontoxic and antigenically inactive, do not induce cancer, and can, in some cases, bond directly with bone without any intervening fibrous–connective tissue layer.

Recent studies on ectopic bone formation (osteoinduction or material-induced osteogenesis) of calcium phosphate biomaterials (with grain sizes less than 100 nm) showed that osteoinduction might be an intrinsic property of calcium phosphate biomaterials. Ripamonti (1991, 1996) reported bone formation in coral-derived hydroxyapatite implanted in muscles of baboons, rabbits, and dogs. Vargervik (1992) reported ectopic bone formation in porous hydroxyapatite ceramics in monkeys; Yamasaki and Saki (1992) found bone formation induced by hydroxyapatite ceramics in dogs; Toth *et al.* (1993) found bone formation in dogs induced by hydroxyapatite–tricalcium phosphate as well as α- and β-calcium pyrophosphate; Klein *et al.* (1994) observed bone tissue in calcium phosphate ceramic in soft tissues of dogs; and Yang *et al.* (1996, 1997) and Yuan *et al.* (1997a–e) reported hydroxyapatite–tricalcium phosphate ceramic-induced osteogenesis in dogs, pigs, and rabbits.

In spite of these investigations, many reports in the literature demonstrate that these nanoapatite ceramics are not always osteoinductive and, furthermore, do not possess mechanical properties similar enough to bone for sustained osseointegration (Muller-Mai *et al.*, 1995; Doremus, 1992; Du *et al.*, 1999; Weng *et al.*, 1997), criteria necessary for increased orthopedic and dental implant efficacy. Moreover, mechanisms of osteoinduction of calcium phosphate ceramics are not clear and seem to depend on specific nanoapatite material properties (such as surface properties and crystallinity) and the animal tested (i.e., dog versus rabbit). Undoubtedly, the incidental cases of calcium phosphate biomaterial-induced osteogenesis indicate promise in the development of nanoapatites with instrinsic osteoinductive properties, but surface properties of these novel prostheses that induce bone growth must obviously be understood.

Surface roughness (as determined by macro- and microporosity, grain size, etc.) crystallinity, and wettability (or hydrophobicity) of calcium phosphate–derived ceramics have all been shown to influence osteoinductivity. For example, hydroxyapatite substrates with high amounts of surface microporosity induced bone formation under the skin of dogs (Yamaski and Saki, 1992). Yuan *et al.* (1999) showed that surfaces with microporosity promoted calcium phosphate ceramic-induced osteogenesis in dorsal muscles of dogs by increasing surface area and subsequent adsorption of proteins and growth factors that stimulated osteoblastic activity. Even the size of macropores influenced osteoinduction; for example, pore diameters of 100 to 600 μm on the surface of hydroxyapatite and calcium phosphate ceramics enhanced bone ingrowth in dogs, rabbits, and humans (Inoue *et al.,* 1992; Kawamura *et al.,* 1987; Flately *et al.,* 1983; Passuti *et al.,* 1989). It has also been suggested that apatite crystals sintered at low temperatures, which possess a low degree of crystallinity, were more active in bone formation (de Bruijn *et al.,* 1994).

In contrast to reports demonstrating the effect of surface properties on the extent of osseointegration of calcium phosphate with juxtaposed bone, few studies have addressed the mechanisms of enhanced osteoblastic activity on these nanoapatites. One set of *in vitro* studies pinpoints grain size in the nanometer regime as the major parameter for enhancing ceramic cytocompatability. For example, compared to respective conventional, larger grain size, ceramic formulations, enhanced adhesion of osteoblasts (the bone-forming cells) and decreased adhesion of fibroblasts (cells that contribute to fibrous encapsulation and callus formation events that may lead to implant loosening and failure) have been observed on nanophase alumina, titania, and HA (Webster *et al.,* 1998, 1999a,b, 2000a). In fact, decreasing alumina grain size from 167 to 24 nm increased osteoblast adhesion 51% and at the same time decreased fibroblast adhesion 235% after 4 hr (Webster *et al.,* 2000a).

Investigations of the underlying mechanism(s) revealed that the concentration, conformation, and bioactivity of vitronectin (a protein contained in serum that is known to mediate osteoblast adhesion ((Thomas *et al.,* 1997); see the section "Vitronectin") was responsible for the select, enhanced adhesion (a crucial prerequisite for subsequent, anchorage-dependent-cell function) of osteoblasts on these novel nanoceramic formulations. Specifically, of the proteins (such as albumin, laminin, fibronectin, collagen, and vitronectin) tested, vitronectin adsorbed in the highest concentration on nanophase alumina after 4 hr; moreover, competitive adsorption of vitronectin was 10% greater on nanophase compared to conventional alumina (Webster *et al.,*

2001a). Furthermore, and in contrast to protein adsorption to conventional ceramics, vitronectin adsorption to nanophase ceramics was controlled by calcium-mediated mechanisms (Webster *et al.*, 2001a). Calcium mediation affected the conformation of vitronectin subsequently adsorbed on nanophase ceramics to promote osteoblast adhesion. Specifically, a novel adaptation of the standard surface-enhanced Raman scattering (SERS) technique provided evidence of increased unfolding of vitronectin adsorbed on nanophase ceramics (Webster *et al.,* 2001a). Protein conformation plays a critical role in mediating subsequent cell interactions (such as cell adhesion on material surfaces; see the section "Protein-Mediated Cell Adhesion on Biomaterial Surfaces"). For example, vitronectin unfolding promoted availability of specific cell-adhesive epitopes (such as arginine–glycine–aspartic acid–serine) for subsequent enhanced osteoblast adhesion; evidence supporting this claim was provided by competitive inhibition studies (Webster *et al.,* 2001a).

Webster *et al.* (2000a) suggested that the topography (such as roughness dictated by nanometer grain size and nm pore size) of nanophase ceramics influenced interactions (such as adsorption and/or configuration or bioactivity) of select proteins that affected subsequent cell adhesion. For example, because of protein stereochemical structure and ceramic pore dimensions, vitronectin (a linear protein 15 nm in length (Ayad *et al.,* 1994); see the section "Vitronectin") preferentially adsorbed to the small (e.g., 0.69, 0.98, and 0.66 nm for alumina, titania, and hydroxyapatite formulations, respectively) pores present in nanophase ceramics, whereas larger proteins that do not promote osteoblast adhesion (such as laminin with a cruciform configuration 70 nm both in length and width (Ayad *et al.,* 1994); see the section "Laminin") adsorbed to the large (e.g., 2.94, 23.3, and 3.1 nm for alumina, titania, and hydroxyapatite, respectively) pores present in conventional ceramics. In addition, variations in ceramic surface topography (such as nanophase and conventional titania with average surface roughness values of 32 nm and 16 nm, respectively (Webster *et al.,* 2000a)) on the same order of magnitude as the size of proteins may have affected adsorbed protein configuration and, thus, availability of bioactive domains (i.e., specific amino acid sequences) that mediate subsequent osteoblast adhesion. Due to protein dimensions in the nanometer regime (see the section "Organic Phase"), through the use of nanoceramics, engineers can now modify a surface to control and manipulate adsorbed protein configuration or bioactivity for increased bone cell interactions; this is, most likely, the largest uninvestigated and promising potential of nanophase ceramics in biomedical applications.

Adhesion of osteoblasts to ceramic surfaces alone, however, is not adequate to achieve long-term osseointegration of orthopedic and dental implants; subsequent osteoblast functions (such as proliferation, synthesis of extracellular matrix proteins, and deposition of extracellular calcium) are required. *In vitro* studies conducted by Webster *et al.* (2000b) also provided the first evidence of enhanced osteoblast proliferation, alkaline phosphatase synthesis, and concentration of extracellular matrix calcium on ceramics of decreased grain size (specifically on nanophase alumina, titania, and hydroxyapatite).

However, other cell functions must also be investigated on proposed biomaterials; this is true because maintainence of healthy bone juxtaposed to an implanted surface requires the formation of a bone-remodeling unit (consisting of activation of bone cells by the action of growth factors such as insulinlike growth factors I and II (Kaplan *et al.,* 1994), resorption of bone by osteoclasts, and formation of new bone by osteoblasts on the site of the old resorbed bone; see the section "Bone Remodeling"). For this reason, Webster *et al.* (2001b) investigated osteoclastlike cell function on nanophase ceramics and demonstrated enhanced synthesis of tartrate-resistant acid phosphatase (TRAP) and subsequent increased formation of resorption pits when ceramic grain size was reduced into the nanometer regime. Such enhanced corresponding events between osteoclasts and osteoblasts may provide a plausible explanation for the observed, improved osseointegration between nanoapatites and juxtaposed bone *in vivo* (Doremus, 1992).

It is imperative to determine properties of nanoceramics that enhance osteoblast and osteoclast functions; Webster *et al.* (2000a,b, 2001a,b) proposed the following:

1. Enhanced osteoblast and osteoclast functions on nanophase ceramics due to increased nanosize surface roughness (Fig. 10). Compared to conventional ceramics, nanophase ceramics possess increased surface roughness (by 35 to 50%) resulting from both decreased grain size and decreased diameter of surface pores (Webster *et al.,* 1999b). These results confirm those obtained by other research groups who reported enhanced functions of osteoblasts and osteoclasts on surfaces with increased roughness (specifically, surfaces with increased roughness values achieved by sandblast, heat treating, acid etching, and machining) (Curtis and Wilkinson, 1997; Degasne *et al.,* 1999; Keller *et al.,* 1994; Gomi *et al.,* 1993; Matsunaga *et al.,* 1999) but are among the first to demonstrate that surface grain sizes of ceramic formulations in the nanometer regime result in nanosize roughness values that selectively promote bone cell functions.
2. Enhanced osteoblast and osteoclast functions on nanophase ceramics due to increased surface wettability. Compared to conventional ceramics, nanophase ceramics exhibit enhanced surface wettability [evidenced, for example, by aqueous contact angles three times smaller

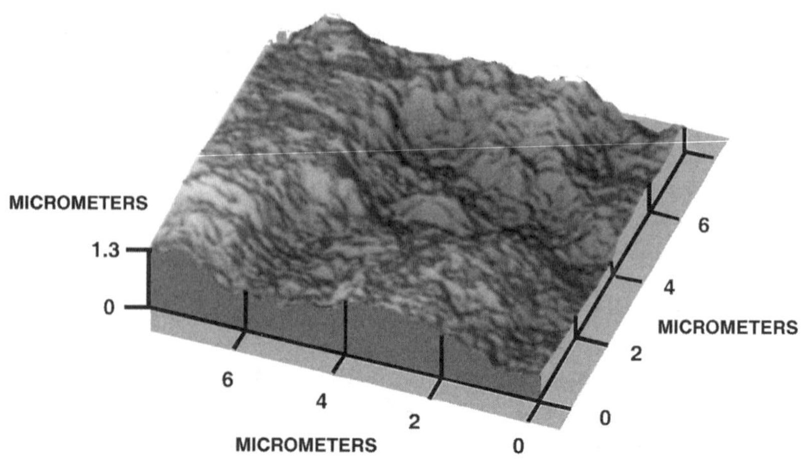

(a) NANOPHASE (39-nm GRAIN SIZE) TITANIA

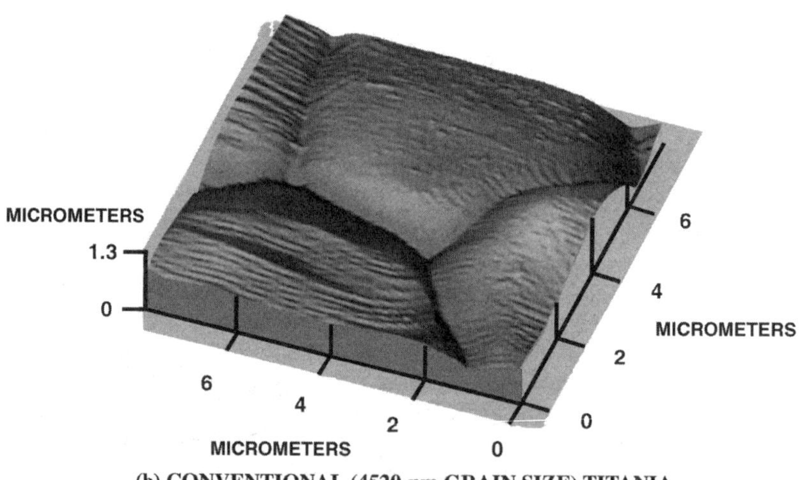

(b) CONVENTIONAL (4520-nm GRAIN SIZE) TITANIA

FIG. 10. Representative topography of nanophase and conventional titania. Representative atomic force micrographs of (a) nanophase titania with 39-nm grain sizes and of (b) conventional titania with 4520-nm grain sizes, illustrating the different topographies of nanophase compared to conventional grain size ceramics.

when alumina grain size was decreased from 167 to 24 nm (Webster *et al.*, 1999b)] due to surface roughness and/or greater numbers of grain boundaries on their surface. These studies confirm results obtained by other research groups who correlated increased adsorption of vitronectin [a protein that promotes osteoblast adhesion (Thomas *et al.*, 1997)] on material surfaces with greater surface wettability (Luck *et al.*, 1998) but are the first to demonstrate that surface grain size of ceramic formulations in the nanometer regime increases the number of grain boundaries at the surface to enhance bioceramic surface wettability for vitronectin adsorption which, subsequently, promotes select bone cell function (Fig. 11).

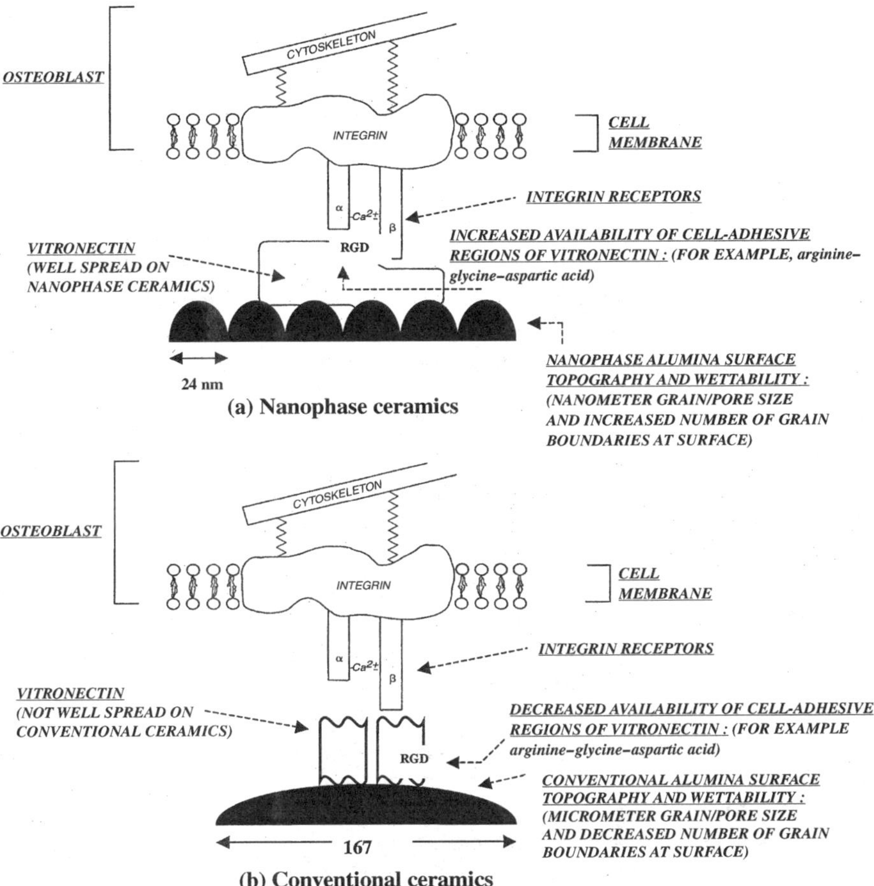

FIG. 11. Unfolding of vitronectin exposes epitopes for osteoblast adhesion on nanophase ceramics. Schematic representation (not in scale) of a possible mechanism for enhanced osteoblast adhesion on (a) nanophase, compared to (b) conventional, ceramics, which involves unfolding of the vitronectin macromolecule to expose select cell-adhesive epitopes (such as arginine–glycine–aspartic acid) for osteoblast adhesion. Increased exposure of cell-adhesive epitopes of vitronectin for enhanced osteoblast adhesion on nanophase ceramics may be due to nanometer surface topography and/or increased wettability due to the greater number of grain boundaries at the surface.

Due to their ability to selectively promote both osteoblast and osteoclast function, nanophase ceramics provide a preferable alternative to conventional orthopedic and dental implants that fail to integrate with surrounding bone; it is undoubtedly highly desirable to minimize, if not avoid, clinical complications that necessitate removal of failed implants as a result of poor surface properties that lead to insufficient osseointegration. These results provide evidence that nanoceramics may be synthesized to match surface properties of bone and, thus, demonstrate strong promise and potential for their use in orthopedic and dental applications.

B. Mechanical Properties of Nanophase Ceramics for Enhanced Orthopedic/Dental Implant Efficacy

1. Rationale

The next generation of orthopedic and dental implants must possess mechanical properties that are similar to those of the surrounding bone. Nanophase ceramics may be synthesized to possess hardness, bending and compressive and tensile strengths that are different than properties of conventional ceramics but similar to those of physiological bone. Indeed, greater mechanical properties (such as hardness, ductility, and enhanced strain to failure) have been reported for ceramics with a reduction in grain size into the nanometer range (Bohn *et al.,* 1991; Mayo *et al.,* 1990, 1992). Mechanical deformation theory indicates that the high-volume fraction of interfacial regions compared to bulk material leads to increased deformation by grain-boundary sliding in nanocrystalline ceramics (Coble, 1963). Similarly, it has been suggested that increased grain-boundary sliding, accompanied by short-range diffusion-healing events as grain size is reduced, results in increased ductility for strongly ionic (that is, ceramic) and covalently bonded nanomaterials (Fig. 12) (Siegel, 1994).

Nanocrystalline materials have been shown to possess increased (by a factor of 2 to 5 for nanocrystalline copper, palladium, and silver (Nieman, 1991; Nieman *et al.,* 1989, 1991a,b)) hardness, to deform at faster (e.g., 34 times faster for 13-nm versus 300-nm zirconia (Ciftcioglu and Mayo, 1990)) superplastic forming rates, to exhibit higher strain rates but without fracture (for example, 40-nm titania deformed without fracture at strains exceeding 0.6 for strain rates as high as 10^3 s^{-1} (Hahn and Averback, 1991; Carry and Mocellin, 1987)), and to bend at lower temperatures than their large-grained (conventional) counterparts (Weertman *et al.,* 1999). For example, Ciftcioglu and Mayo (1990) demonstrated a fourfold reduction in grain size of yittria-stabilized zirconia-accelerated strain rates close to the deformation stress and strain rates typically used for formation of metals via superplastic techniques.

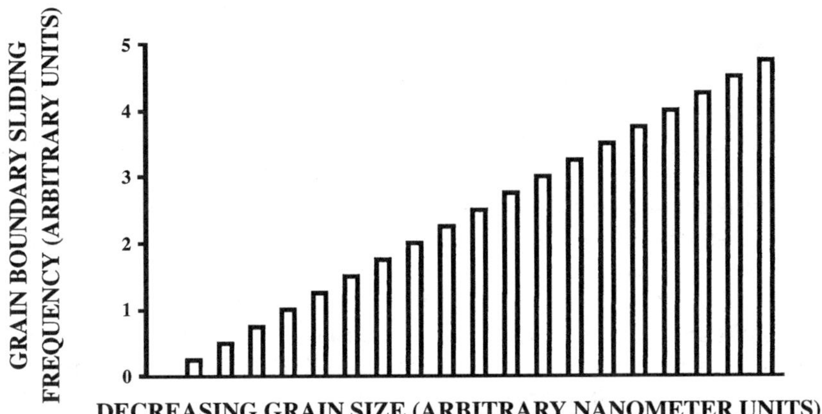

FIG. 12. Schematic deformation properties of nanophase ceramics. For ceramics, decreasing the grain size into the nanometer regime corresponds to an increase in grain boundary sliding, resulting in increased ductility of these materials. (Adapted and redrawn from Siegel, 1994.)

Extended strain to failure has been well documented for submicrometer and nanocrystalline ceramics; elongations to failure of 100 to 800% have been reported for nanophase ceramics whose conventional grain-size counterparts normally fail at 0 to 2% elongation (Maehara and Langdon, 1990; Nieh *et al.,* 1991). Weertman *et al.* (1999) reported plastic deformations of about 100% in compression of single calcium fluoride crystals at temperature conditions under which conventional materials would fail in the elastic regime. True compressive creep testing of nanoceramic materials has substantiated enhanced plasticity; compressive creep tests conducted at moderate temperatures on 99% dense nanophase TiO_2 showed extensive deformation without crack formation (Hahn and Averback, 1991). Nanophase TiO_2 also exhibited enhanced plasticity in tension; Cui and Hahn (1992) subjected 40-nm grain size TiO_2 to biaxial bulge tests at temperatures between 700 and 800°C and reported ductile tensile behavior, without formation of any cracks, as the samples deformed up to true strain levels of 0.1 (Cui and Hahn, 1992). Considering that the brittle nature of conventional, larger-grain-size, ceramics inhibits their widespread use as orthopedic or dental materials, it is interesting to ponder if and how the enhanced mechanical properties of nanophase ceramics could be incorporated into the next generation of biomaterials with improved osseointegrative capabilities.

2. *Experimental Evidence*

To date, few research groups have incorporated the enhanced mechanical properties of nanophase ceramics into orthopedic and dental applications.

One example in the literature demonstrates that nanophase (10–15 nm average grain size) diamond-coated (via microwave plasma chemical vapor deposition) titanium implants exhibited improved fracture toughness and adhesion compared to conventional coatings (Catledge and Vohra, 1999). Toprani *et al.* (2000) reported levels of strain more indicative of ductile coatings than ceramic during indentation tests. The authors note that the observed increased fracture toughness comes at the expense of decreased hardness, which may be improved by synthesizing layers of nanocrystalline diamond on layers of high phase purity microcrystalline diamond (Catledge *et al.*, 2000); by controlling placement of nano- and microcrystalline layers, one can obtain a compositelike film whose toughness, hardness, and surface roughness properties can be tailored as desired. For this reason, future biomaterials consisting of nanophase ceramics can be tailored to meet clinical requirements associated with anatomical differences or patient age; such requirements arise because, for example, the modulus of elasticity varies by 10% in bone from the human hip and tibia and by up to 10% in bone from 10- and 90-year-old human (Kaplan *et al.*, 1994). Conventional orthopedic and dental implants are not synthesized to simulate differences in mechanical properties of bone based on anatomical differences or patient age; instead, a "one-size-fits-all" model is currently used.

Another example in the literature demonstrating how mechanical properties of nanophase ceramics could enhance orthopedic and dental implants is by Webster *et al.* (1999a), who demonstrated that compared to respective conventional ceramics, bending properties (specifically, bending modulus, bending strength, flexural rigidity, and bending structural stiffness) of nanophase alumina, titania, and hydroxyapatite ceramics are closer to those of human femur bone; for example, the average bending modulus of nanophase alumina (namely, 35 GPa) was 1.8 times greater, whereas conventional alumina was 2.7 times greater (specifically, 52 GPa) than the respective value of human femur bone (in the range of 19 GPa) (Fung, 1993). Moreover, compared to values of the bending strength for conventional hydroxyapatite (in the range 38–113 MPa), Ahn *et al.* (2000) also showed bending strengths of nanocrystalline hydroxyapatite (182 MPa) closer to those of physiological bone (160 MPa).

Undoubtedly, changes in porosity (such as bulk and surface porosity and diameter of individual pores) of nanophase ceramic formulated by Webster *et al.* (1999) provide an explanation for the observed differences in mechanical properties of respective nanophase and conventional ceramic formulations; for example, individual surface pores four times smaller were achieved in nanophase (67-nm grain size) compared to conventional (179-nm grain size) HA formulations (Webster *et al.*, 2000a). Compared to conventional formulations, therefore, the bending modulus of nanophase ceramics may be the result of the combined effects of (1) increased grain-boundary sliding mechanisms due to a larger number of grain boundaries present in nanophase ceramics, and/or (2) smaller crack initiation sites due to smaller diameters of individual surface pores (i.e., fewer and less surface flaws) present on nanophase ceramics. For these reasons, nanophase ceramics provide a preferable alternative to conventional orthopedic and dental implants that fail due to crack initiation and propagation during *in vivo* loading; it is, undoubtedly, highly desirable to minimize, if not avoid, clinical complications that necessitate removal of failed implants as a result of poor mechanical properties. These results provided evidence that nanophase ceramics may be synthesized to match bending properties of bone and thus demonstrated strong promise and potential for their use in orthopedic and dental implant applications.

VII. Conclusions

Nanostructured ceramics provide alternatives not yet fully explored for orthopedic and dental implant applications; the improved mechanical properties of these novel ceramic formulations, in addition to their established exceptional biocompatibility, constitute characteristics that promise improved orthopedic and dental efficacy. Requirements applicable for the design of nanophase ceramics for orthopedic and dental applications include the following:

1. Mechanical properties (such as bending, hardness, and compressive and tensile strength) of bioceramics similar to human bone can be obtained by decreasing the grain size of ceramic formulations into the nanometer regime. Such mechanical properties must be incorporated into bioceramics for orthopedic and dental applications; mechanical properties similar to those of physiological bone are needed in order to minimize imbalances in stress and strain distributions at the tissue–implant interface, which often lead to bone resorption (i.e., osteolysis) and eventual implant loosening and failure.
2. Surface properties (such as topography and wettability) of bioceramics similar to human bone can be obtained by decreasing the grain size of ceramic formulations into the nanometer regime. Such surface properties must be incorporated into proactive bioceramics for orthopedic and dental applications; surface properties similar to those of physiological bone are needed in order to promote select cell interactions that lead to sufficient osseointegration between an orthopedic or

dental implant and juxtaposed bone. Sufficient bonding of juxtaposed bone to an implanted surface stabilizes the prostheses *in situ,* minimizes motion-induced damage to surrounding tissues, and is crucial to the clinical success of orthopedic and dental implants.

A proactive orthopedic or dental bioceramic should be designed to promote the adsorption of vitronectin, a protein that optimizes subsequent adhesion of osteoblasts. Besides adsorption, proactive bioceramics for orthopedic and dental applications should promote unfolding and thus enhance the bioactivity of vitronectin by exposing epitopes (such as integrin and heparan sulfate-binding sites) necessary for select osteoblast adhesion. Unfolding and increased bioactivity of vitronectin can be accomplished by nanophase ceramic topography (as controlled by nanometer surface grain size) and increased wettability (due to higher numbers of grain boundaries at the surface).

As the disciplines of cell–tissue engineering and nanophase material science develop and mature, the preceding design criteria will be expanded and refined. Undoubtedly, nanophase ceramics have great potential to become the next generation of choice proactive biomaterials for innovative biotechnology and biomedical applications that could have profound clinical impact.

REFERENCES

Ahn, E., Gleason, N. J., Nakahira, A., and Ying, J. Y., Properties of nanostructured hydroxyapatite-based bioceramics. *Proc. Sixth World Biomaterials Congress.* 643 (2000).

Anderson, J. M., Mechanisms of inflammation and infection with implanted devices. *Cardiovasc. Pathol.* **2,** 33S–41S (1993).

Ayad, S., Boot-Handford, R., Humphries, M. J., Kadler, K. E., and Shuttleworth, A., "The Extracellular Matrix Factsbook." Academic Press, San Diego, 1994, pp. 29–149.

Baraton, M. I., Chen, X., and Gonsalves, K. E., FTIR study of nanostructured aluminum nitride powder surface: determination of the acidic/basic sites by CO, CO_2 and acetic acid adsorptions. *Nanostructured Materials* **8** (4), 435–445 (1997).

Beer, F., and Johnston, E. R., "Mechanics of Materials." McGraw-Hill Book Company, New York, 1981, p. 585.

Bohn, R., Haubold, R., Birringer, R., and Gleiter, H., Nanocrystalline intermetallic compounds—An approach to ductility. *Scripta Metall. Mater.* **25,** 811 (1991).

Brunette, P. M., The effect of surface topography on cell migration and adhesion, *in* "Surface Characterization of Biomaterials: Progress in Biomedical Engineering, Volume 6" (B. D. Ratner, Ed.), pp. 203–217. Elsevier, New York, 1988.

Brunski, J. B., Influence of biomechanical factors at the bone-biomaterial interface, *in* "The Bone-Biomaterial Interface" (J. E. Davies, Ed.), pp. 391–404, University of Toronto Press, Toronto, 1991.

Carry, C., and Mocellin, A., Structural superplasticity in single phase crystalline ceramics *Ceramics International* **13** (2), 89–98 (1987).

Catledge, S., Baker, P., Tarvin, J., and Vohra, Y., Multilayer nanocrystalline/microcrystalline diamond films studied by laser reflectance interferometry, *in* Diamond and Related Mater., **9** (8), 1512–1517 (2000).

Catledge, S., and Vohra, Y., Effect of nitrogen addition on the microstructure and mechanical properties of diamond films grown using high-methane concentrations. *J. App. Phys.* **86** (1), 698–700 (1999).

Ciftcioglu, M., and Mayo, M. J., Processing of nanocrystalline ceramics, *in* "Superplasticity in Metals, Ceramics and Intermetallics Symposium Proceedings" (M. J. Mayo, M. Kobayashi, J. Wadsworth, Eds.), pp. 77–86. Materials Research Society, Pittsburgh (1990).

Coble, R. L., Development of microstructure in ceramic systems. *J. of App. Phy.* **34,** 1679 (1963).

Cox, D., Aoki, T., Seki, J., Motoyama, Y., and Yoshida, K., The pharmacology of the integrins. *Medical Research Reviews* **14,** 195–228 (1994).

Cui, Z., and Hahn, H., Tensile deformation of nanostructured TiO_2 at low temperatures. *Nanostructured Materials* **1,** 419 (1992).

Curtis, A., and Wilkinson, C., Review: Topographical control of cells. *Biomaterials* **18** (24), 1573–1583 (1997).

Dalton, B. A., McFarland, C. D., Gengenbach, T. R., Griesser, H. J., and Steele, J. G., Polymer surface chemistry and bone cell migration. *J. Biomat. Sci. Polym. Ed.* **9** (8), 781–799 (1995).

Darnell, J., Lodish, H., and Baltimore, D., "Molecular Cell Biology." W.H. Freeman and Company, New York, 1990.

Davies, J. E., The importance and measurement of surface charge species in cell behaviour at the biomaterial interface, *in* "Surface Characterization of Biomaterials: Progress in Biomedical Engineering, Volume 6" (B. D. Ratner, Ed.), pp. 219–234. Elsevier, New York, 1988.

Davies, J. E., and Matsuda, T., Extracellular matrix production by osteoblasts on bioactive substrates *in vitro. Scanning Microscopy* **2,** 1445–1452 (1994).

de Bruijn, J., Bovell, Y. P., and van Blitterswijk, C., Osteoblast and osteoclast responses to calcium phosphates. *Bioceramics* **7,** 293–298 (1994).

Dee, K. C., Considerations for the design of proactive dental/orthopaedic implant biomaterials, Ph.D. thesis, Rensselaer Polytechnic Institute, 1996.

Dee, K. C., Andersen, T. T., Rueger, D. C., and Bizios, R., Conditions which promote mineralization at the bone/implant interface: a model *in vitro* study. *Biomaterials* **17,** 209–215 (1996).

Degasne, I., Basle, M. F., Demais, V., Hure, G., Lesourd, M., Grolleau, B., Mercier, L., and Chappard, D., Effects of roughness, fibronectin and vitronectin on attachment, spreading, and proliferation of human osteoblast-like cells (Saos-2) on titanium surfaces. *Calcified Tissue International* **64** (6), 499–507 (1999).

Doremus, R. H., Review: bioceramics. *J. Mat. Sci.* **27,** 285–297 (1992).

Du, C., Cui, F. Z., Zhu, X. D., and de Groot, K., Three-dimensional nano-Hap/collagen matrix loading with osteogenic cells in organ culture. *J. Biomed. Mat. Res.* **44,** 407–415 (1999).

Ducheyne, P., Bioactive ceramics. *J. Bone Joint Surg.* **76B,** 861–862 (1994).

Ducheyne, P., Stimulations of biological function with bioactive glass. *MRS Bull.* **23** (11)**,** 43–49 (1999).

Ellingsen, J. E., A study on the mechanism of protein adsorption to TiO_2. *Biomaterials* **12** (6), 593 (1991).

Flately, T. J., Lynch, K. L., and Benson, M., Tissue response to implants of calcium phosphate ceramics in the rabbit spine. *Clinical Orthop.* **179,** 246–252 (1983).

Fung, Y. C., "Biomechanics Mechanical Properties of Living Tissues Second Edition." Springer-Verlag, New York, 1993, pp. 500–538.

Garcia, R., and Doremus, R. H., Electron microscopy of the bone-hydroxyapatite interface from a human dental implant. *J. Mat. Sci.: Materials in Medicine* **3,** 154–156 (1992).

Garvey, B. T., and Bizios, R., A transmission electron microscopy examination of the interface between osteoblasts and metal biomaterials. *J. Biomed. Mat. Res.* **29** (8), 987–992 (1995).

Gehron-Robey, P., The biochemistry of bone. *Endocrinology and Metabolism Clinics of North America* **18,** 859–902 (1989).

Gomi, K., Lowenberg, B., Shapiro, G., and Davies, J. E., Resorption of sintered hydroxyapatite by osteoclasts *in vitro*. *Biomaterials* **14** (2), 91–96 (1993).

Graf, J., Ogle, R. C., Robey, F. A., Sasaki, M., Martin, G. R., Yamada, Y., and Kleinman, H. K., A pentapeptide from the laminin B1 chain mediates cell adhesion and binds the 67000 laminin receptor. *Biochemistry* **26,** 6896 (1987).

Grenoble, D. E., Katz, J. L., Dunn, K. L., Gilmore, R. S., and Murty, K. L., The elastic properties of hard tissues and apatites. *J. Biomed. Mat. Res.* **6,** 221–233 (1972).

Gronowicz, G., and McCarthy, M. B., Response of human osteoblasts to implant materials: integrin-mediated adhesion. *J. Orthop. Res.* **14** (6), 878–887 (1996).

Hahn, H. J., and Averback, R. S., Low-temperature creep of nanocystalline titanium (IV) oxide. *J. Amer. Ceramic Soc.* **74** (11), 2918–2921 (1991).

Healy, K. E., Lom, B., and Hockberger, P. E., Spatial distribution of mammalian cells dictated by material surface chemistry. *Biotech. Bioeng.* **43,** 792–800 (1994).

Heegard, A., Structure and molecular regulation of bone matrix proteins. *J. Bone Mineral Res.* **8,** S843–S847 (1993).

Hench, L. L., Bioceramics: from concept to clinic. *Amer. Ceramic Soc. Bull.* **72,** 93–98 (1993).

Hench, L. L., and Ethridge, E. C., Biomaterial—The interfacial problem. *Adv. Biomed. Eng.* **5,** 35–150 (1975).

Horbett, T. A., Principles underlying the role of adsorbed plasma proteins in blood interactions with foreign materials. *Cardiovasc. Pathol.* **2,** 137S–148S (1993).

Horbett, T. A., Proteins: Structure, Properties and Adsorption to Surfaces. In "Biomaterials Science: An Introduction to Materials in Medicine," (B. D. Ratner, A. S. Hoffman, A. S. Schoen, and J. E. Lemons, Eds.), pp. 133–140. Academic Press, New York, 1996.

Hynes, R. O., Integrins: versatility, modulation, and signaling in cell adhesion. *Cell* **69,** 11–25 (1992).

Inoue, O., Shimabukura, H., Shingaki, Y., and Ibaraki, K., Our application of high porosity hydroxyapatite cubes for the treatment of non-cystic benign tumors. *Bioceramics* **5,** 411–418 (1992).

Izzard, C. S., Radinsky, R., and Culp, L. A., Substratum contacts and cytoskeletal reorganization of BALB/c3T3 cells on a cell-binding fragment and heparin-binding fragments of plasma fibronectin. *Experimental Cell Research* **165,** 320–336 (1986).

Jarcho, M., Kay, J. F., Gumaer, K. I., Doremus, R. H., and Drobeck, H. P., Tissue, cellular and subcellular events at a bone-ceramic hydroxylapatite interface. *J. Bioeng.* **1,** 79–92 (1977).

Kaplan, F. S., Hayes, W. C., Keaveny, T. M., Boskey, A., Einhorn, T. A., and Iannotti, J. P., Form and function of bone, *in* "Orthopaedic Basic Science" (S. P. Simon, Ed.), pp. 127–185. American Academy of Orthopaedic Surgeons, Columbus, Ohio, 1994.

Kasemo, B., and Lausmaa, J., Surface science aspects of inorganic biomaterials. *C.R.C. Critical Rev. in Biocompatibility* **2,** 335–380 (1986).

Kawamura, M., Iwata, H., and Miura, T., Chondroosteogenic response to crude bone matrix proteins bound to hydroxyapatite. *Clinical Orthop.* **217,** 281–292 (1987).

Keaveny, T. M., and Hayes, W. C., Mechanical properties of cortical and trabecular bone. *Bone* **7,** 285–344 (1993).

Keller, J. C., Stanford, C. M., Wightsman, J. P., Draughn, R. A., and Zaharias, R., Characterization of titanium implant surfaces. III. *J. Biomed. Mat. Res.* **28,** 939–946 (1994).

Klabunde, K. J., Stark, J., Koper, O., Mohs, C., Park, D., Decker, S., Jiang, Y., Lagadic, I., and Zhang, D., Nanocrystals as stoichiometric reagents with unique surface chemistry. *J. Phys. Chem.* **100** (30), 12142–12153 (1996).

Klein, C., de Groot, K., Chen, W., Li, Y., and Zhang, X., Osseous substance formation in porous calcium phosphate ceramics in soft tissues. *Biomaterials* **15,** 31–34 (1994).

Kramer, R. H., Enenstein, J., and Waleh, N. S., Integrin structure and ligand specificity in cell-matrix interactions, *in* "Molecular and Cellular Aspects of Basement Membranes" (D. H. Rohrbach and R. Timpl, Eds.), pp. 239–258. Academic Press, New York, 1993.

Laterra, J., Silbert, J. E., and Culp, L. A., Cell surface heparan sulfate mediates some adhesive responses to glycosaminoglycan-binding matrices including fibronectin. *J. Cell Biol.* **96,** 112–121 (1983).

Lausmaa, J., Kasemo, B., Matsson, H., and Odelius, H., Multi-technique surface characterizations of oxide films on electropolished and anodically oxidized titanium *App. Surf. Sci.* **45,** 189–200 (1990).

Lehman, W. B., Strongwater, A. B., Tunc, D., Kummer, F., Atar, D., Grant, A. D., Kramer, M., and Rohovsky, M. W., Internal fixation with biodegradable plate and screw in dogs. *J. Pediatric. Orth.,* Part B **3,** 190–193 (1994).

Li, P., Kanasniemi, I., and de Groot, K., Bonelike hydroxyapatite induction by a gel-derived titania on a titanium substrate. *J. Am. Ceram. Soc.* **77,** 1307–1312 (1994).

Lopes, M. A., Monteiro, F. J., Santos, J. D., Serro, A. P., and Saramago, B., Hydrophobicity, surface tension, and zeta potential measurements of glass-reinforced hydroxyapatite composites. *J. Biomed. Mater. Res.* **45** (4), 370–375 (1999).

Luck, M., Paulke, B.-R., Schroder, W., Blunk, T., and Muller, R. H., Analysis of plasma protein adsorption on polymeric nanoparticles with different surface characteristics. *J. Biomed. Mat. Res.* **39,** 478–485 (1998).

Maehara, Y., and Langdon, T. G., Superplasticity in ceramics. *J. Mater. Sci.* **25,** 2275 (1990).

Malik, M. A., Puleo, D. A., Bizios, R., and Doremus, R. H., Osteoblasts on hydroxyapatite, alumina and bone surfaces *in vitro:* Morphology during the first 2 h of attachment *Biomaterials* **13** (2), 123–128 (1992).

Martin, B. R., and Burr, D. B., "Structure Function and Adaptation of Compact Bone." Raven Press, New York, 1989.

Mathews, C. K., and van Holde, K. E., "Biochemistry." Benjamin/Cummings, Redwood City, CA, 1990.

Matsunaga, T., Inoue, H., Kojo, T., Hatano, K., Tsujisawa, T., Uchiyama, C., and Uchida, Y., Disaggregated osteoclasts increase in resorption activity in response to roughness of bone surface. *J. Biomed. Mat. Res.* **48** (4), 417–423 (1999).

Mayo, M., Siegel, R. W., Liao, Y. X., and Nix, W. D., Nanoindentation of nanocrystalline ZnO. *J. Mat. Res.* **7,** 973 (1992).

Mayo, M., Siegel, R. W., Narayanasamy, A., and Nix, W. D., Mechanical properties of TiO_2 as determined by nanoindentation. *J. Mat. Res.* **5,** 1073 (1990).

Moursi, A. M., Damsky, C. H., Lull, J., Zimmerman, D., Doty, S. B., Aota, S-I., and Globus, R. K., Fibronectin regulates calvarial osteoblast differentiation. *J. Cell Sci.* **109,** 1369–1380 (1996).

Moursi, A. M., Globus, R. K., and Damsky, C. H., Interactions between integrin receptors and fibronectin are required for calvarial osteoblast differentiation *in vitro*. *J. Cell Sci.* **110,** 2187–2196 (1997).

Muller-Mai, C. M., Stupp, S. I., Voigt, C., and Gross, U., Nanoapatite and organoapatite implants in bone: Histology and ultrastructure of the interface. *J. Biomed. Mater. Res.* **29,** 9–18 (1995).

Nakamura, H., and Ozawa, H., Immunohistochemical localization of heparan sulfate proteoglycan in rat tibiae. *J. Bone Mineral Res.* **9,** 1289–1299 (1994).

Nieh, T. G., Wadsworth, J., and Wkai, F., Recent advances in superplastic ceramics and ceramic composites. *Int. Mater. Rev.* **36,** 146 (1991).

Nieman, G. W., Processing and mechanical behavior of nanocrystalline Cu, Pd and Ag., Ph.D. Thesis, Northwestern University, 1991.

Nieman, G. W., Weertman, J. R., and Siegel, R. W., Microhardness of nanocrystalline palladium and copper produced by inert gas condensation. *Scripta Metallurgica* **23,** 2013 (1989).

Nieman, G. W., Weertman, J. R., and Siegel, R. W., Mechanical behavior of nanocrystalline metals. *J. Mat. Res.* **6,** 1012 (1991a).

Nieman, G. W., Weertman, J. R., and Siegel, R. W., Mechanical behavior of Nanocrystalline Cu and Pd, *in* "Microcomposites and Nanophase Materials" (D. C. Van Aken, Ed.), p. 15. TMS, Warrendale, 1991b.

Park, J. B., and Lakes, R. S., "Biomaterials: an Introduction Second Edition," Plenum Press, New York, 1992, pp. 79–244.

Passuti, N., Daculsi, G., Rogez, J. M., Martin, S., and Bainvel, J. V., Macroporous calcium phosphate ceramic performance in human spine fusion. *Clinical Orthop.* **248,** 169–176 (1989).

Pereira, M. M., Clark, A. E., and Hench, L. L., Calcium phosphate formation on sol–gel-derived bioactive glasses *in vitro*. *J. Biomed. Mater. Res.* **28,** 693–698 (1994).

Praemer, A., Furner, S., and Rice, S. D., "Musculoskeletal Conditions in the United States." American Academy of Orthopaedic Surgery, Park Ridge, IL, 1992.

Puleo, D. A., and Bizios, R., Mechanisms of fibronectin-mediated attachment of osteoblasts to substrates *in vitro*. *Bone and Mineral* **18,** 215–226 (1992).

Puleo, D. A., Preston, K. E., Shaffer, J. B., and Bizios, R., Examination of osteoblast-orthopaedic biomaterials interactions using molecular techniques. *Biomaterials* **14,** 111–114 (1993).

Radin, S. R., and Ducheyne, P., The effect of calcium phosphate ceramic composition and structure on *in vitro* behavior. II. Precipitation. *J. Bone and Mineral Res.* **27,** 35–45 (1993).

Ratner, B., New ideas in biomaterials science—A path to engineered biomaterials. Society for Biomaterials 1992 Presidential Address. *J. Biomed. Mat. Res.* **27,** 837–850 (1992).

Rifkin, B. R., and Gay, C. V., "Biology and Physiology of the Osteoclast." Academic Press, Boca Raton, FL, 1992.

Ripamonti, U., The morphologenisis of bone in replicas of porous hydroxapatite obtained from conversion of calcium carbonate exoskeletons of coral. *J. Bone Jt. Surg.* **73A,** 692–703 (1991).

Ripamonti, U., Osteoinduction in porous hydroxyapatite implanted in heterotopic sites of different animal models. *Biomaterials* **17,** 31–35 (1996).

Schakenraad, J. M., Cells: their surface and interactions with materials, *in* "Biomaterials Science: An Introduction to Materials in Medicine" (B. D. Ratner, A. S. Hoffman, A. S. Schoen, and J. E. Lemons, Eds.), pp. 141–146. Academic Press, New York, 1996.

Schneider, G., and Burridge, K., Formation of focal adhesions by osteoblasts adhering to different substrata. *Exp. Cell. Res.* **214** (1), 264–269 (1994).

Schwartz, M. A., Transmembrane signaling by integrins. *Trends Cell Biol.* **2,** 304–308 (1992).

Siegel, R. W., Nanophase materials, *in* "Encyclopaedia of Applied Physics," Vol. 11., 173–199, VCH Publishers, New York, 1994.

Siegel, R. W., Creating nanophase materials. *Sci. Amer.* **275,** 42–47 (1996).

Siegel, R. W., and Fougere, G. E., Mechanical properties of nanophase materials, *in* "Nanophase Materials: Synthesis-Properties-Applications" (G. C. Hadjipanayis and R. W. Siegel, Eds.), p. 233. Kulwer, Dordrecht, 1994.

Siegel, R. W., and Fougere, G. E., Mechanical properties of nanophase metals. *Nanostructured Materials* **6,** 205 (1995a).

Siegel, R. W., and Fougere, G. E., Grain size dependent mechanical properties in nanophase materials. *Material Research Society Symposium Proc.* **362,** 219 (1995b).

Sinha, R. K., and Tuan, R. S., Regulation of human osteoblast integrin expression by orthopaedic implant metals. *Bone* **18,** 451–457 (1996).

Steele, J. G., McFarland, C., Dalton, B. A., Johnson, G., Evans, M. D. M., Howlett, C. R., and Underwood, P. A., Attachment of human derived bone cells to tissue culture polystyrene and to unmodified polystyrene: The effect of surface chemistry upon initial cell attachment. *J. Biomat. Sci. Polymer Ed.* **5,** 245–257 (1993).

Stein, G. S., and Lian, J. B., Molecular mechanisms mediating proliferation/differentiation interrelationships during progressive development of the osteoblast phenotype. *Endocrine Reviews* **14,** 424–442 (1993).

Stein, G. S., Lian, J. B., and Owen, T. A., Relationship of cell growth to the regulation of tissue-specific gene expression gene expression during osteoblast differentiation. *FASEB J.* **4,** 3111–3123 (1990).

Thomas, C. H., McFarland, C. D., Jenkins, M. L., Rezania, A., Steele, J. G., and Healy, K. E., The role of vitronectin in the attachment and spatial distribution of bone derived cells on materials with patterned surface chemistry. *J. Biomed. Mater. Res.* **37,** 81–93 (1997).

Tiller, J., Berlin, P., and Klemm, D., A novel efficient enzyme-immobilization reaction on NH_2 polymers by means of L-ascorbic acid. *Biotechnol. Appl. Biochem.* **30** (2), 155–162 (1999).

Toprani, N., Catledge, S., and Vohra, Y., Interfacial adhesion and toughness of nanostructured diamond coatings. *J. Mater. Sci.* **15** (5), 1052–1055 (2000).

Toth, J. M., Lynch, K. L., and Hackbarth, D. A., Ceramic-induced osteogenesis following subcutaneous implantation of calcium phosphates. *Bioceramics* **6,** 9–13 (1993).

Trippel, S. B., Potential role of insulinlike growth factors in fracture healing. *Clinical Orthopaedics* **355S,** S301–313 (1998).

Turkova, J., Oriented immobilization of biologically active proteins as a tool for revealing protein interactions and function. *J. Chromatogr. B Biomed. Sci. Appl.* **722** (1–2), 11–31 (1999).

Underwood, P. A., and Bennett, F. A., A comparison of the biological activities of the cell-adhesive proteins vitronectin and fibronectin. *J. Cell Sci.* **93** (4), 641–649 (1989).

Vargervik, K., Critical sites for new bone formation, *in* "Bone Grafts and Bone Substitutes" (M. B. Habal and A. H. Reddi, Eds.), p. 112–120. W. B. Saunders, Philadelphia, 1992.

Webster, T. J., Siegel, R. W., and Bizios, R., An *in vitro* evaluation of nanophase alumina for orthopaedic/dental applications, *in* "Bioceramics 11: 11th International Symposium on Ceramics in Medicine" (R. Z. LeGeros and J. P. LeGeros, Eds.), p. 273–276. World Scientific, New York, 1998.

Webster, T. J., Siegel, R. W., and Bizios, R., Design and evaluation of nanophase alumina for orthopaedic/dental applications. *Nanostructured Mat.* **12,** 983–986 (1999a).

Webster, T. J., Siegel, R. W., and Bizios, R., Osteoblast adhesion on nanophase ceramics. *Biomaterials* **20,** 1221–1227 (1999b).

Webster, T. J., Ergun, C., Doremus, R. H., Siegel, R. W., and Bizios, R., Specific proteins mediate enhanced osteoblast adhesion on nanophase ceramics. *J. Biomed. Mat. Res.* **51** (3), 475–483 (2000a).

Webster, T. J., Ergun, C., Doremus, R. H., Siegel, R. W., and Bizios, R., Enhanced functions of osteoblasts on nanophase ceramics. *Biomaterials* **21,** 1803–1810 (2000b).

Webster, T. J., Schadler, L. S., Siegel, R. W., and Bizios, R., Mechanisms of enhanced osteoblast adhesion on nanophase alumina involve vitronectin. *Tissue Engineering* **7** (3), 291–302 (2001a).

Webster, T. J., Ergun, C., Doremus, R. H., Siegel, R. W., and Bizios, R., Enhanced functions of osteoclast-like cells on nanophase ceramics. *Biomaterials* **22** (11), 1327–1333 (2001b).

Weertman, J. R., Farkas, D., Hemker, K., Kung, H., Mayo, M., Mitra, R., and van Swygenhoven, H., Structure and mechanical behavior of bulk nanocrystalline materials. *MRS Bulletin* **24** (2), 44–50 (1999).

Wen, X., Wang, X., and Zhang, N., Microrough surface of metallic biomaterials: a literature review. *Biomed. Mat. Eng.* **6** (3), 173–189 (1996).

Weng, J., Liu, Q., Wolke, J. G. C., Zhang, X., and de Groot, K., Formation and characteristics of the apatite layer on plama-sprayed hydroxyapatite coatings in simulated body fluid. *Biomaterials* **18** (15), 1027–1035 (1997).

Yamada, H., *in* "Strength of Biological Materials" (F. G. Evans, Trans.). Williams and Wilkins, Baltimore, MD, 1970.

Yamasaki, H., and Saki, H., Osteogenic response to porous hydroxyapatite ceramics under the skin of dogs. *Biomaterials* **13,** 308–312 (1992).

Yang, Z., Yuan, H., Tong, W., Zou, P., Chen, W., and Zhang, X., Osteogenesis in extraskeletally implanted porous calcium phosphate ceramics: Variability among different kinds of animals. *Biomaterials* **17,** 2131–2137 (1996).

Yang, Z., Yuan, H., Zou, P., Tong, W., Qu, S., and Zhang, X., Osteogenic responses to extraskeletally implanted synthetic calcium phosphate ceramics an early stage histomorphological study in dogs. *J. Mater. Sci. Med.* **8,** 697–701 (1997).

Yuan, H., Yang, Z., Zou, P., Li, Y., and Zhang, X., Rapid osteogenesis in porous biphasic calcium phosphate ceramics implanted in domestic pigs. *Biomed. Eng. Appl. Bas. Com.* **9,** 268–273 (1997a).

Yuan, H., Li, Y., Yang, Z., Feng, J., and Zhang, X., An investigation on the osteoinduction of synthetic porous phase-pure hydroxyapatite ceramic. *Biomed. Eng. Appl. Bas. Com.* **9,** 274–278 (1997b).

Yuan, H., Li, Y., Yang, Z., Feng, J., and Zhang, X., Calcium phosphate ceramic induced osteogenesis in rabbits, *in* "Biomedical Materials Research in the Far East (III)," (X. Zhang and Y. Ikada, Eds.), pp. 228–229. Kobunshi Kankokai, Kyoto, Japan, 1997c.

Yuan, H., Li, Y., Yang, Z., and Zhang, X., Osteoinduction of pure β-TCP ceramic in dogs, *in* "Biomedical Materials Research in the Far East (III)," (X. Zhang and Y. Ikada, Eds.), pp. 188–189. Kobunshi Kankokai, Kyoto, Japan, 1997d.

Yuan, H., Li, Y., Kurashina, K., and Zhang, X., "Host tissue response of calcium phosphate cement," *in* Biomedical Materials Research in the Far East (III). (X. Zhang and Y. Ikada, Eds.), Kobunshi Kankokai, Kyoto, Japan, 1997e, p. 116–117.

Yuan, H., Kurashina, K., de Bruijn, J., Li, Y., de Groot, K., and Zhang, X., A preliminary study on osteoinduction of two kinds of calcium phosphate ceramics. *Biomaterials* **20,** 1799–1806 (1999).

Yubao, L., Klein, C. P. A. T., Xingdong, Z., and de Groot, K., Formation of bone apatite-like layer on the surface of porous hydroxyapatite ceramics. *Biomaterials* **15,** 835–841 (1994).

FABRICATION, STRUCTURE, AND TRANSPORT PROPERTIES OF NANOWIRES

Yu-Ming Lin,[1] Mildred S. Dresselhaus,[1,2] and Jackie Y. Ying[3]*

[1]Department of Electrical Engineering and Computer Science, [2]Department of Physics, and [3]Department of Chemical Engineering, Massachusetts Institute of Technology, Cambridge Massachusetts 02139

I. Introduction 168
II. Fabrication and Structural Characteristics of Nanowires 168
A. Template-Assisted Synthesis 169
B. Laser-Assisted Synthesis 181
C. Other Synthesis Methods 184
III. Theoretical Modeling of Nanowire Band Structures 185
A. Band Structures of One-Dimensional Systems 185
B. The Semimetal–Semiconductor Transition in Semimetallic Nanowires 188
IV. Transport Properties 191
A. Semiclassical Model 192
B. Temperature-Dependent Resistivity of Nanowires 193
V. Summary 198
References 199

Nanowire systems have attracted a great deal of attention recently due to their technological potential. They are of fundamental interest because they exhibit unique quantum confinement effects. In this article, advances in the fabrication of nanowires via template-assisted and laser-assisted approaches are reviewed. The structure and characteristics of different nanowire systems are discussed. To understand and predict the unusual properties of nanowires, we have developed a generalized theoretical model for the band structure of these one-dimensional systems. A unique semimetal–semiconductor transition that occurs in bismuth nanowires is described. Transport measurements on bismuth and antimony nanowires illustrate that these novel materials are very different from their bulk counterparts. A transport

* To whom correspondence should be addressed.

ADVANCES IN CHEMICAL ENGINEERING, VOL. 27
0065-2377/01 $35.00

model, based on the band-structure calculations, is presented to explain the experimental results and to gain insight into the transport phenomena of nanowire systems.

I. Introduction

Nanostructured materials have received significant attention in recent years because of their fundamental importance and potential applications in areas ranging from chemistry, physics, biology, and materials science. In the fields of electronics and optics, the drive to miniaturize devices and increase storage density has fueled research in nanotechnology. Several techniques have been developed to fabricate nanostructures, such as epitaxial growth, electron beam lithography, chemical vapor deposition, and self-assembly approaches. Nanostructures represent a new class of materials with properties different from molecular species and bulk solid-state structures. They exhibit quantum confinement effects, giving rise to unique behavior that can be exploited in novel optical, magnetic, electronic, and thermoelectric devices. Multiple quantum well structures are probably the most studied nanostructured systems; the carriers are confined to two dimensions in these systems. In comparison, one-dimensional quantum wires and zero-dimensional quantum dots are expected to show even stronger quantum confinement effects. Quantum wires are perhaps the most amenable for the design of novel electronic devices because they exhibit more pronounced quantum effects than the two-dimensional structures; unlike most zero-dimensional systems, they maintain transport continuity along the wire axis. This chapter describes the fabrication, structure, and transport properties of nanowire systems of interest to electronic applications. It presents a generalized theoretical model for the band structure of quantum wires, which is responsible for the novel behavior of these systems. In addition, a semiclassical transport model is also developed so that the unusual transport properties that are experimentally observed in nanowire systems can be explained and compared to the predicted behavior.

II. Fabrication and Structural Characteristics of Nanowires

The preparation of one-dimensional quantum systems represents one of the greatest challenges in materials fabrication. The synthesis of highly crystalline and continuous nanowires is essential for studying the interesting quantum phenomena in these low-dimensional systems. Over the past decade, significant progress has been made in deriving nanowires via techniques, such as high-pressure injection (Huber *et al.*, 1994, 1999; Zhang *et al.*, 1998a, 1999), vapor deposition (Heremans *et al.*, 2000; Cheng *et al.*, 1999), electrochemical deposition (Foss *et al.*, 1992; Hornyak *et al.*, 1997; Fasol, 1998; Piraux *et al.*, 1994, 1999; Martin, 1994; Sun *et al.*, 1999a; Whitney *et al.*, 1993; Blondel *et al.*, 1994; Liu *et al.*, 1998a; Routkevitch *et al.*, 1996a,b; Yi and Schwarzacher, 1999; Peng *et al.*, 2000; Zeng *et al.*, 2000), laser ablation (Morales and Lieber, 1998; Yu *et al.*, 1998; Zhang *et al.*, 1998b, 2000a; Duan *et al.*, 2000), thermal evaporation (Wang *et al.*, 1998a; Tang *et al.*, 1999), molecular beam epitaxy (Nötzel *et al.*, 1992; Omi and Ogino, 1997), and electron beam lithography (Chou *et al.*, 1996). These different methods led to the generation of a broad range of one-dimensional nanostructured materials (Martin, 1994). Although some approaches are more suitable than others for specific applications, each approach may present certain synthetic limitations with regard to nanowire diameter, crystallinity, fabrication costs, and scalability. In this section, the nanowire-fabrication techniques will be reviewed, and the structure of the nanowires produced will be discussed.

A. Template-Assisted Synthesis

The template-assisted synthesis of nanowires is a conceptually elegant way of fabricating nanostructures in an organized assembly (Ozin, 1992; Tonucci *et al.*, 1992; Ying, 1999). The template is typically a nonconducting host matrix containing nanometer-sized elongated pores or voids, which are filled by the material of choice that adopts the pore morphology. If the pore diameters are sufficiently small, the pore-filling nanowires would exhibit quantum confinement effects. In this synthesis approach, the important factors for consideration include the template characteristics, such as chemical stability, mechanical properties, pore diameter, uniformity, and density. Templates that have been used for nanowire synthesis include anodic alumina, nanochannel glass, ion track-etched polymers, and mica films.

Anodic alumina templates are produced by the anodic oxidation of aluminum films in acidic electrolyte solutions (Diggle *et al.*, 1969; O'Sullivan and Wood, 1970; Li *et al.*, 1998a). The resulting aluminum oxide film possesses a regular hexagonal array of parallel and nearly cylindrical channels (Keller *et al.*, 1953; Masuda *et al.*, 1997; Li *et al.*, 1998a), as depicted in Fig. 1. Depending on the anodization conditions, such as the voltage applied and the nature and concentration of electrolyte used, the pores of anodic alumina can be systematically varied from less than 10 nm to 200 nm in diameter,

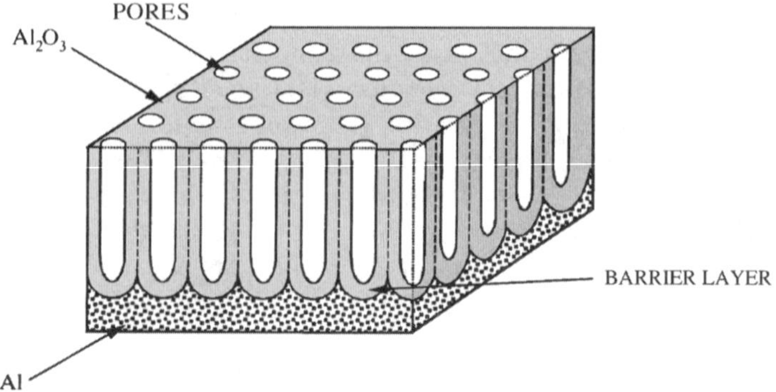

FIG. 1. Schematic of a porous anodic alumina template.

with a packing density on the order of 10^9 to $10^{11}/cm^2$ (Diggle *et al.,* 1969; O'Sullivan and Wood, 1970; AlMawlawi *et al.,* 1991; Zhang *et al.,* 1999). The film thickness (or pore length) ($< 1\ \mu m$ to $> 50\ \mu m$) can be controlled by the duration of the anodization process. With their synthetic flexibility, ordered pore structure, high pore density, and high mechanical strength, anodic alumina systems have gained great popularity as a template material for nanowire fabrication. In addition, with recent improvements on the degree of pore ordering, anodic alumina has stimulated much interest for other device applications (Govyadinov and Zakhvitcevich, 1998; Davydov *et al.,* 1999; Li *et al.,* 1999; Masuda *et al.,* 1999; Li *et al.,* 2000; Kouklin *et al.,* 2000) because it provides a highly ordered two-dimensional nanopattern, readily and inexpensively, on a large scale compared to conventional lithographic approaches.

In the processing of anodic alumina films (Keller *et al.,* 1953; Masuda *et al.,* 1997; Li *et al.,* 1998a; Zhang *et al.,* 1999), a thin aluminum sheet is first mechanically and electrochemically polished to produce a smooth surface. It is then anodized in an acidic solution at a constant voltage and temperature. The anodization voltage V determines the interpore distance D by the empirical relation $D\,(\text{nm}) = -1.7 + 2.81 \cdot V\,(\text{volts})$ (Li *et al.,* 1998a). Different electrolytes are usually used for different anodization voltage ranges: 20 wt% sulfuric acid (H_2SO_4) for less than 20 V, 4 wt% oxalic acid ($H_2C_2O_4$) for 30–65 V, and 3.5 wt% phosphoric acid (H_3PO_4) for 70 V or more. The as-prepared anodic alumina film has open pores on the top surface of the substrate and is capped by a barrier layer on the other side (see Fig. 1). Therefore, the sample has to be etched by an acidic solution to remove the barrier layer for suitable applications (Li *et al.,* 1998a; Jessensky *et al.,* 1998). Figures 2(a) and (b) show scanning electron microscopy (SEM) images of the top surfaces of porous alumina templates anodized in 4 wt% oxalic acid and 20 wt% sulfuric acid, respectively. The pore ordering and uniformity of these materials have been optimized by a two-step anodization technique (Li *et al.,* 1998b; Lin *et al.,* 2000).

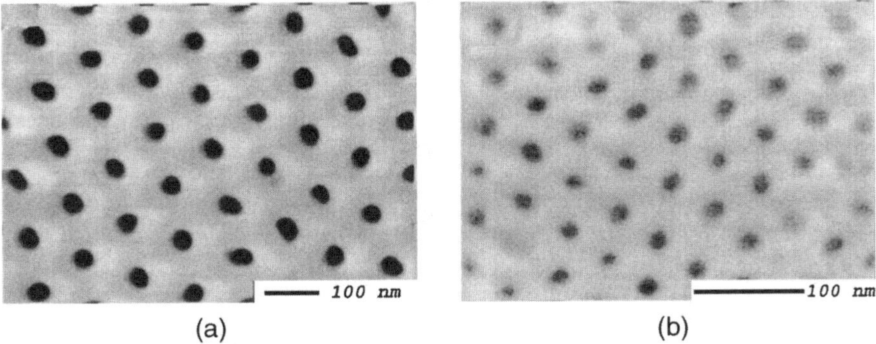

FIG. 2. SEM images of the top surfaces of porous anodic alumina templates anodized in (a) 4 wt% $H_2C_2O_4$ and (b) 20 wt% H_2SO_4. The average pore diameters in (a) and (b) are 44 nm and 18 nm, respectively.

The self-organized pore structure in anodic alumina is derived from two coupled processes, pore formation and pore ordering. Pore formation is generally believed to be a result of several mechanisms, including oxide formation and dissolution. During the anodization process, anions (O^{2-} or OH^-) migrate through the oxide layer and form Al_2O_3 at the oxide–metal interface, whereas some of the Al^{3+} ions produced at the oxide–metal interface move through the oxide layer and become ejected into the electrolyte. The nonuniform electric field and current density present in the sample from surface topological variations are essential to the pore growth mechanism. The surface variations may come from either the initial sample polishing or the self-induced variation during steady-state pore growth. The field-enhanced dissolution or the increased local temperature promotes the dissolution rate in some areas at the oxide-electrolyte interface to yield the observed pore morphology. The self-ordering of the pores may be attributed to the volume expansion during the oxide formation, which produces a repulsive force between the pores. In response to the mutual repulsive forces between neighboring pores, self-arrangements of the cylindrical pores will occur to maximize their packing density in the regular hexagonal order at steady state.

Porous templates can also be fabricated by chemically etching particle tracks originated from ion bombardment (Ferain and Legras, 1993; Sun *et al.*, 1999a). The pores produced by this track-etching method are randomly distributed with a packing density of 10^7 to $10^9/cm^2$, which is substantially lower than that in anodic alumina. By controlling the duration of chemical etching, the pore diameter can be varied from several hundred nanometers to about 5 nm. Track-etched polycarbonate membranes are commercially available in a variety of pore sizes and have been widely used in the template-assisted synthesis of nanowires (Martin, 1994; Blondel *et al.*, 1994; Liu *et al.*, 1998a).

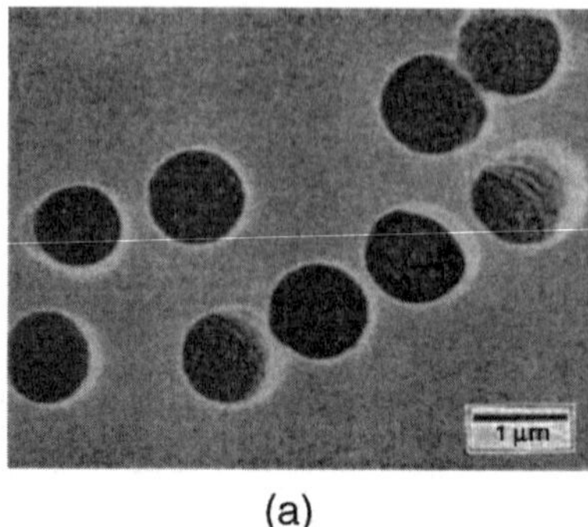

(a)

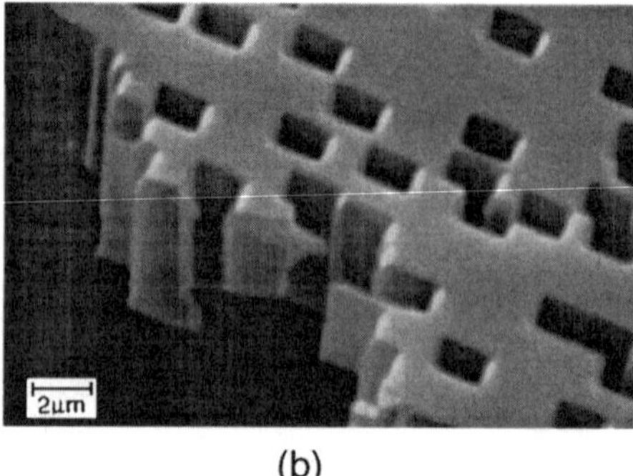

(b)

FIG. 3. (a) SEM image of the particle track-etched polycarbonate membrane, with a pore diameter of 1 μm (Martin, 1994). (b) SEM image of 2-μm pores in a single-crystal mica film prepared by particle track-etching (Sun *et al.*, 2000a).

Recently, single-crystalline mica films have also been used for the fabrication of track-etched pores (Sun *et al.*, 1999a). Polymer membranes have the disadvantage that they are relatively soft, so the definition of the pore morphology is not straightforward, and their internal pore surfaces can be quite rough. Also, they are fairly limited in working temperatures for sample fabrication and characterization compared to mica films, which are chemically stable up to 770 K. Figures 3(a) and (b) show the SEM images of etched tracks in polycarbonate membranes (Martin, 1994) and mica films (Sun *et al.*, 2000), respectively. The track-etched pores in the polymeric membranes are approximately cylindrical. Due to anisotropic etching rates, the pores in mica films have a diamond-shaped cross section with a tapered pore wall along the ion track (Sun *et al.*, 1999a).

Nanochannel glass (NCG) has also been proposed for the template-assisted synthesis of nanowires (Tonucci *et al.*, 1992; Huber *et al.*, 1994). It contains a regular hexagonal array of capillaries similar to the pore structure in anodic alumina. It is prepared by arranging two dissimilar glasses in a predetermined configuration. For example, Fig. 4(a) shows the use of cylindrical rods of an acid-etchable glass as the cores in a hexagonally packed matrix of inert glass tubes. The array of this core-tube assembly is then drawn at high temperatures to reduce the cross-sectional area. By repeating the drawing process, the channel diameters can be made as small as 33 nm, with a packing density of 3×10^{10} pores/cm^2 (Tonucci *et al.*, 1992). Finally, the core glass rods are etched away by an acid, yielding an array of cylindrical pores. Figure 4(b) shows a SEM micrograph of an NCG with 33-nm channels arranged in a hexagonal close packing (Tonucci *et al.*, 1992).

A number of other porous materials may be used as the host matrices for nanowire fabrication. A mesoporous molecular sieve termed MCM-41 (Beck *et al.*, 1992) possesses hexagonally packed pores with very small

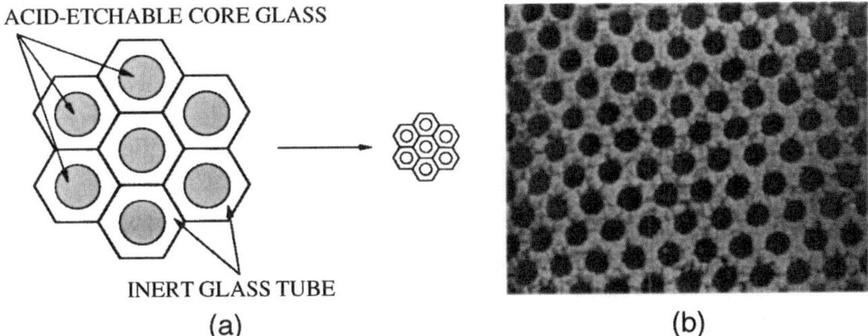

FIG. 4. (a) Schematic diagram illustrating the fabrication process for nanochannel glass array (Tonucci *et al.*, 1992). (b) SEM image of a glass array with 33-nm channels after acid etching (Tonucci *et al.*, 1992).

channel diameters, which can be systematically varied between 2 nm and 10 nm using a surfactant-based supramolecular template (Ying *et al.*, 1999). The surfactant molecule consists of a hydrophilic head group that has a high affinity for water and a long hydrophobic tail group. In the presence of water, surfactant species will undergo self-aggregation to form micellar structures, where the hydrophilic head groups are directed outward in contact with water, whereas the hydrophobic tail groups form the micellar core to minimize contact with water. Silicate precursors are then deposited on the ordered arrays of micelles, which are self-assembled at a high enough micellar concentration to form inorganic–organic mesostructures. The organic surfactants can be removed via heat treatment to yield silicates with hexagonally packed cylindrical pores. Wu and Bein (1994) have fabricated conducting organic filaments in the nanochannels of MCM-41. Han *et al.* (2000) have also prepared metallic nanowires in SBA-15 silica with hexagonally packed mesopores templated with triblock copolymers. Recently, the DNA molecule was also used as a template for growing nanometer-sized wires (Braun *et al.*, 1998).

The variety of porous solid materials that can be used as templates for nanostructure synthesis has been reviewed by Ozin (1992). In the template-assisted synthesis of nanowires, the pores or voids of the template are filled with the chosen material using a number of approaches. Nanowires have been derived via pressure injection, electrochemical deposition, and vapor deposition, as described in the following sections.

1. *Pressure Injection*

In the pressure injection method, the nanowires are formed by injecting the desired material in its liquid form into the pores of the template. This technique has been used to fabricate a number of metallic and

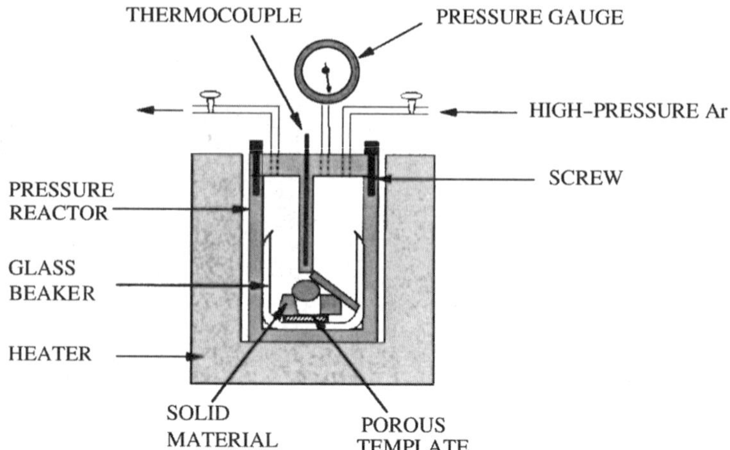

FIG. 5. Schematic of the experimental setup for the pressure injection of materials into the nanochannels of a porous template.

semiconducting nanowires (Huber *et al.*, 1994; Zhang *et al.*, 1998a; Lin *et al.*, 2000b). Before the injection process, the template is first cleaned to remove any particles or substances that may prevent the pores from being filled. The template is then placed with the solid material in a high-pressure chamber. The chamber is evacuated at a temperature slightly below the melting point of the solid material for a few hours to degas the template. It is then brought to a temperature above the melting point of the solid, so that the porous template is immersed in a liquid melt. Next, the chamber is disconnected from the vacuum system and filled with an inert gas such as argon. The high pressure within the chamber forces the liquid melt into the pores of the template. Following this pressure injection process, the chamber is slowly cooled to solidify the impregnated material within the nanochannels of the template before releasing the inert gas. By carefully controlling the cooling rate for nanowire solidification, it is possible to fabricate essentially single-crystalline wires (Zhang *et al.*, 1999). Figure 5 shows a schematic diagram of the experimental setup for the pressure injection process. The detailed pressure injection procedures are described by Huber *et al.* (1994) and Zhang *et al.* (1998a).

In the pressure injection process, a greater pressure is needed to overcome the surface tension for the liquid melt to enter pores of a smaller diameter. The relation between the pore diameter d_w and the required pressure P is given by the Washburn equation (Adamson, 1982),

$$d_W = -4\gamma \cos\theta/P, \tag{1}$$

where γ is the surface tension of the liquid and θ is the contact angle between the liquid and the template. For most liquid metals and semiconductors of interest, the surface tension ranges from 100 to 600 dyne/cm. Assuming the least favorable case of a nonwetting interface ($\theta = 180^\circ$) and a medium surface tension of 400 dyne/cm, filling channels 40 nm in diameter would require a pressure of about 400 bar. To reduce the pressure required or to maximize the filling factor, additives may be used to decrease the surface tension or the contact angle. For example, it was found that the introduction of a small quantity of copper atoms into the bismuth melt could facilitate the injection of liquid bismuth into porous anodic alumina (Zhang *et al.,* 1999).

For the pressure injection approach, the templates employed have to be chemically stable and structurally robust under the high pressures and temperatures involved. Anodic alumina and nanochannel glass can be used as template materials for this nanowire synthesis approach. Metallic (e.g. Bi, In, Sn and Al) and semiconducting (Se, Te, GaSb and Bi_2Te_3) nanowires have been pressure injected into anodic alumina templates (Huber *et al.,* 1994; Zhang *et al.,* 1998a; Lin *et al.,* 2000b). The following describes some of the important properties of bismuth nanowires fabricated by the pressure injection method.

Figure 6(a) shows a SEM image of bismuth nanowires embedded in an anodic alumina template after pressure injection. The pore diameter of the template was about 42 nm, and the maximum pressure and temperature applied for the injection process were about 310 bar and 325°C, respectively. A small amount of copper was introduced into liquid bismuth to enhance its injection; the copper atoms should be segregated from bismuth during solidification because they have zero solubility in solid bismuth. Upon solidification, the copper flakes were brought to the top surface of Bi due to their

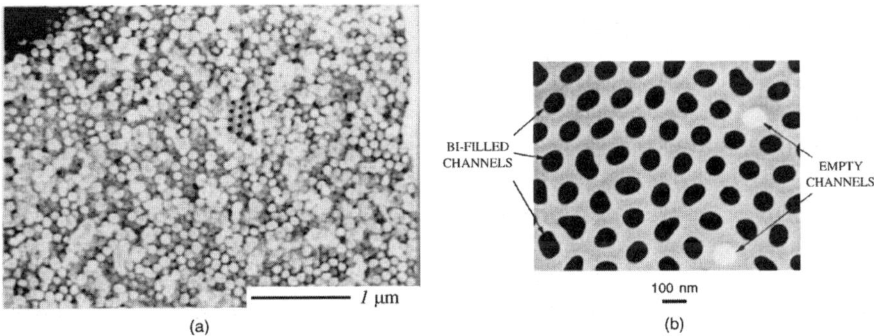

FIG. 6. (a) SEM image of the bottom surface of an anodic alumina template filled with bismuth. The pore diameter is 42 nm. (b) TEM micrograph of the cross section of a 65-nm bismuth nanowire array (Zhang *et al.,* 1999).

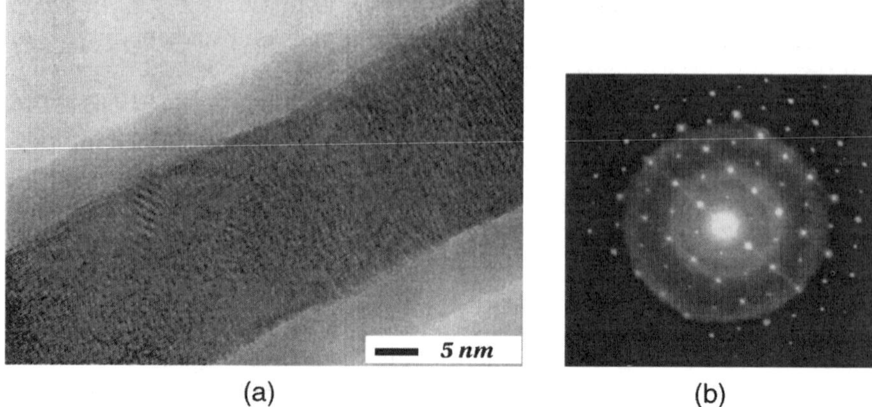

FIG. 7. (a) A HRTEM image of a 40-nm freestanding bismuth nanowire, showing lattice fringes. The amorphous surface layer is bismuth oxide formed upon air exposure of bismuth nanowire. (b) SAED pattern of a single Bi nanowire (Zhang *et al.,* 1999).

lower density, leaving pure bismuth nanowires within the pore channels. Figure 6(b) shows a TEM image of the cross section of a 65-nm Bi nanowire array, illustrating a high pore-filling factor (Zhang *et al.,* 1999). Because Bi has a higher electron density than anodic alumina, it is shown in dark contrast in the TEM image. Figure 7(a) shows a high-resolution transmission electron microscopy (HRTEM) image of a 40-nm freestanding bismuth nanowire, which was obtained by dissolving the anodic alumina template in a special acid solution that did not attack the bismuth nanowires. The freestanding wire was found to have a nearly uniform diameter (within 10%) along its length. The lattice fringes in Fig. 7(a) indicated a highly crystalline structure, as confirmed by the selected-area electron diffraction (SAED) pattern shown in Fig. 7(b). It was found that the alumina matrix protected the bismuth nanowires against oxidation, whereas freestanding wires were gradually oxidized upon exposure to air. The amorphous surface layer in Fig. 7(a) is the bismuth oxide developed when the nanowire was exposed to air prior to imaging.

Figure 8 shows X-ray diffraction (XRD) patterns of bismuth nanowire arrays (Lin *et al.,* 2000b). It illustrates that the crystal structure of bismuth nanowires is the same as that of bulk bismuth and that no copper phases were present. The nanowires have a preferred wire orientation dependent on their diameters. The major orientations of the 95-nm and 40-nm bismuth nanowire arrays were normal to the (202) and (012) lattice planes, respectively, indicating that most ($> 80\%$) of the nanowires were oriented along the $[10\bar{1}1]$ and $[01\bar{1}2]$ directions for $d_w \geq 60\,\text{nm}$ and $d_w \leq 50\,\text{nm}$, respectively (Zhang *et al.,* 1999; Lin *et al.,* 2000b). The existence of more than one dominant orientation in the 52-nm Bi nanowires (Fig. 8(b)) was

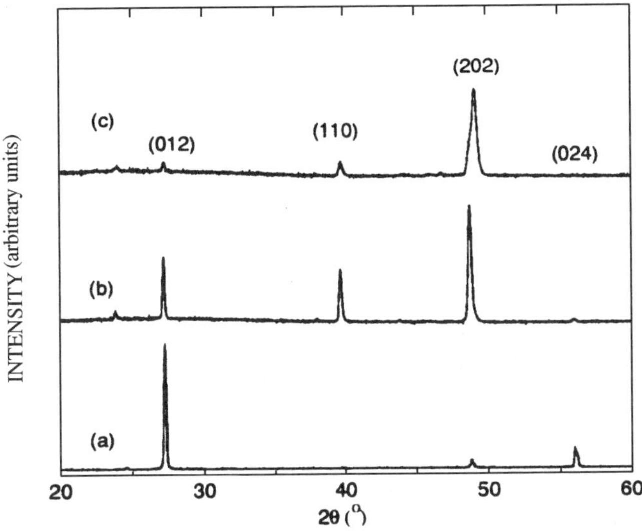

FIG. 8. XRD patterns of bismuth–anodic alumina nanocomposites with average bismuth wire diameters of (a) 40 nm, (b) 52 nm, and (c) 95 nm (Lin *et al.*, 2000b). The Miller indices corresponding to the lattice planes of bulk Bi are indicated above the individual peaks.

attributed to the transitional behavior of nanowires of 'intermediate' diameters as the preferential growth orientation was shifted from $[10\bar{1}1]$ to $[01\bar{1}2]$ with decreasing d_w. Huber *et al.* (2000) found a preferred crystal orientation [0001] along the wire axes for bismuth nanowires fabricated at a much higher pressure (about 1.5 kbar), indicating the possibility that the wire growth orientation might also depend on the applied pressure and other process parameters during the injection process. Additional XRD studies also revealed the presence of a metastable high-stress phase in the as-prepared bismuth nanowires (Zhang *et al.*, 1999), which could be converted to the normal phase by a thermal annealing treatment.

The smallest diameter attained for bismuth nanowires by the pressure injection method was about 13 nm, using a pressure of approximately 0.3 kbar (Zhang *et al.*, 1999). Finer nanowires might be fabricated by increasing injection pressures (Huber *et al.*, 2000), but it remains to be seen if the anodic alumina templates would retain their structural integrity under those high pressures.

2. *Electrochemical Deposition*

Electrochemical deposition has attracted increasing attention as a technique for nanowire fabrication. Traditionally, electrochemistry has been used to grow thin films on conducting surfaces. Because electrochemical deposition is usually controllable in the direction normal to the substrate surface, this method can be readily extended to fabricate one-dimensional or zero-dimensional nanostructures if the deposition is confined within the pores of an appropriate template. The electrochemical technique involves first coating a thin conducting metal film onto one side of the porous membrane to serve as the cathode for electroplating. The length of the nanowires deposited can be controlled by the duration of the electroplating process. Electrochemical deposition has been used to synthesize nanowires and superlattices (Piraux *et al.*, 1994; Blondel *et al.*, 1994) of a variety of materials, including metals (e.g. Bi (Piraux *et al.*, 1999; Liu *et al.*, 1998a), Co (Ferre *et al.*, 1997; Zeng *et al.*, 2000), Fe (AlMawlawi *et al.*, 1991; Peng *et al.*, 2000), Ni (Ferre *et al.*, 1997; Sun *et al.*, 1999a), Cu (Piraux *et al.*, 1994; Blondel *et al.*, 1994), Ag (Bhattacharrya *et al.*, 2000), and Au (Hornyak *et al.*, 1997)), semiconductors (e.g., CdS (Routkevitch *et al.*, 1996a; Kouklin *et al.*, 2000)), superconductors (e.g., Pb (Yi and Schwarzacher, 1999)), and conducting polymers (Martin, 1994; Piraux *et al.*, 1999).

For electrochemical deposition, it is critical to choose a template system that is chemically stable in the electrolyte used and during the electrolysis process. The presence of cracks and defects in the templates will affect the electrochemical process, because deposition will occur predominantly in the more accessible cracks, leaving most of the nanochannels unfilled. Particle track-etched mica films or polymer membranes have been employed in simple dc electrolysis. To use anodic alumina in dc electrochemical deposition, the insulating barrier layer has to be removed first, and a metal film has to be evaporated onto one side of the template. However, the rectifying properties of the oxide barrier layer make it possible to use the ac deposition method directly. Although the applied voltage is sinusoidal and symmetrical, the current is greater during the cathodic half-cycles, so that deposition is dominant over the stripping that occurs in the subsequent anodic half-cycles. By retaining the barrier layer and using the ac electrolysis approach, the problem with cracks may be avoided, because there is no rectification at the defects, and the deposition process is reversed during each anodic half-cycle. By this approach, metals (e.g., Co (Zeng *et al.*, 2000) and Fe (AlMawlawi *et al.*, 1991; Peng *et al.*, 2000)) and semiconductors (e.g., CdS (Routkevitch *et al.*, 1996a; Kouklin *et al.*, 2000)) have been deposited within the pores of anodic alumina templates without removing the barrier layer.

In contrast to nanowires synthesized by the pressure injection method, nanowires derived electrochemically are typically polycrystalline with no preferred crystal orientation. However, some exceptions occur. For example, polycrystalline CdS nanowires fabricated by ac electrodeposition in anodic alumina were shown to have a preferred wire growth orientation along the c axis (Routkevitch *et al.*, 1996). Recently, Xu *et al.* (2000a, 2000b) prepared

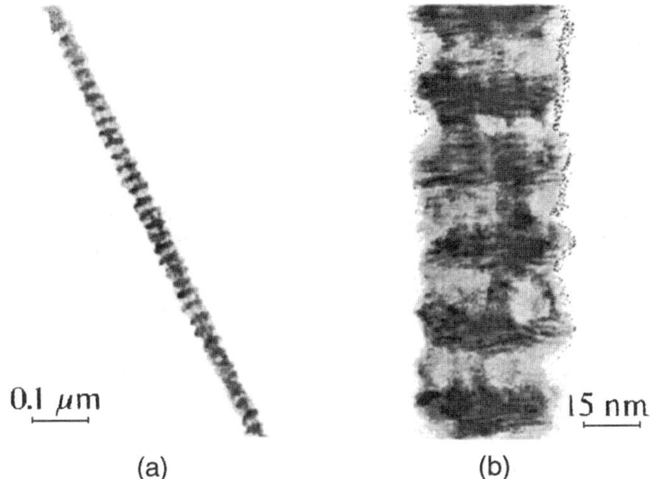

FIG. 9. (a) TEM image of a single Co(10 nm)/Cu(10 nm) multilayered nanowire. (b) A selected region of the sample at high magnification. (Piraux *et al.*, 1994.)

a number of single-crystalline nanowires of II–VI semiconductors (e.g., CdS, CdSe, and CdTe) in anodic alumina by dc electrodeposition in a nonaqueous electrolyte. Yi and Schwarzacher (1999) found that single-crystalline Pb nanowires can be formed by pulse electrodeposition under an overpotential, but no specific crystal orientation was noted along the wire axis.

One advantage of the electrochemical deposition technique is the possibility of fabricating multilayered structures within the nanowire systems. By varying the cathodic potentials in the electrolyte that contains two different kinds of ions, different metal layers are deposited controllably. Co–Cu multilayered nanowires have been synthesized to study the giant magnetoresistance (GMR) effect (Piraux *et al.*, 1994; Blondel *et al.*, 1994). Figure 9 shows TEM images of a single Co/Cu nanowire that is approximately 40 nm in diameter prepared by Piraux *et al.* (1994). The light bands represent Co-rich regions and the dark bands represent Cu-rich layers. This electrodeposition method provides a low-cost, effective approach for preparing multilayered one-dimensional nanostructures.

3. *Vapor Deposition*

Vapor deposition methods include physical vapor deposition (PVD) (Heremans *et al.*, 2000) and chemical vapor deposition (CVD) (Cheng *et al.*, 2000) or metallorganic chemical vapor deposition (MOCVD) (Berry *et al.*, 1996). Vapor deposition and electrochemical deposition are capable of deriving nanowires of smaller diameters ($\leq$ 20 nm) more readily than pressure injection methods because they do not involve the use of high pressures to deposit materials inside the porous channels.

The experimental setup and procedures for the physical vapor deposition of bismuth nanowires are described by Heremans *et al.* (2000). The target material is placed in a crucible that is covered with the porous template to be filled. The template used for vapor deposition has open pores on both sides. The crucible, surrounded by heating wires, is inserted into a vacuum chamber. It is heated to melt the target material and to provide a high vapor pressure. Because the pressure outside is lower than that within the crucible, the vapor passes through the pores of the template. A slow cooling process is then initiated to produce a temperature gradient across the porous template. Because the temperature of the outer surface of the template is lower than that of the inner surface, the vapor begins to condense within the pores from the outer surface, and nanowires are grown inward. The process is completed once the temperature on the inner surface of the template falls below the melting point of the target material. Heremans *et al.* (2000) have synthesized nearly single-crystalline bismuth nanowires within anodic alumina by this approach; these nanowires possess a preferred crystal growth orientation along the wire axis, similar to that prepared by the pressure injection method (Zhang *et al.*, 1999; Heremans *et al.*, 2000).

Compound materials can also be prepared by the vapor deposition technique with two reacting gases (Cheng *et al.*, 2000). The reacting species with the lower melting point is placed in the crucible. The crucible, covered by the porous template, is inserted into a tube that contains the other reacting gas. The tube is heated up to vaporize the reactants in the crucible. The two gases then react to produce compound wires within the nanochannels of the template. Single-crystalline GaN nanowires have been synthesized in an anodic alumina template through reaction of Ga_2O vapor with a flowing ammonia stream (Cheng *et al.*, 1999, 2000). A different liquid/gas-phase approach has been used by Berry *et al.* (1996) to prepare polycrystalline GaAs and InAs nanowires in a nanochannel glass array. In this case, the nanochannels are filled with one liquid precursor (e.g., trimethyl gallium or triethyl indium) via a capillary effect, and the nanowires are formed by reaction between the liquid precursor and the gas reactant (e.g., AsH_3).

Recently, carbon nanotubes, an important class of one-dimensional nanostructures, have been fabricated within the pores of anodic alumina via CVD (Davydov *et al.*, 1999; Li *et al.*, 1999; Iwasaki *et al.*, 1999; Suh *et al.*, 1999). A small amount of metal (e.g., Co) is first electrochemically deposited on the bottom of the pores as a catalyst for the carbon nanotube growth, and the template is heated to 700 to 800°C in a flowing gas mixture of N_2 and acetylene or ethylene. The hydrocarbon molecules are then pyrolyzed to form carbon nanotubes within the porous template. Using anodic alumina as the matrix, a highly ordered two-dimensional array of carbon nanotubes has been achieved. Such well-aligned nanotube array has stimulated much interest with its great potential for applications such as cold-cathode flat-panel displays.

B. Laser-Assisted Synthesis

In addition to the template-assisted method, laser-assisted synthesis has been developed for generating nanowires. It is particularly useful for producing large quantities of crystalline semiconducting nanowires with ultrafine diameters ($\leq$ 10 nm). Nanowire synthesis by the laser-assisted method is based on the vapor-liquid-solid (VLS) growth of single-crystalline silicon whiskers (Wagner and Ellis, 1964), which was discovered in the early 1960s. In VLS growth, a liquid metal droplet or catalyst cluster forms an energetically favored site for the adsorption of gas-phase reactants and the nucleation site for crystallization when supersaturated. The reactant crystallizes at the surface of the liquid cluster, and a preferentially one-dimensional structure is developed. Figure 10 shows a schematic of the silicon nanowire growth by the VLS mechanism. The one-dimensional structure obtained by such growth has a diameter larger than 0.1 μm, which is limited by the minimum size of the liquid droplet.

Recently, a laser ablation–condensation technique was used to produce nanometer-sized catalyst clusters to grow nanowires by the VLS method. A schematic of the laser ablation apparatus used by Morales and Lieber (1998) to produce silicon nanowires is shown in Fig. 11. The target consists of silicon and the catalyst material (e.g., $Si_{1-x}Fe_x$), and a pulsed laser is used to produce nanometer-sized catalyst clusters within a reaction chamber at 1200°C. The ablated materials are carried by an argon gas flow, and the

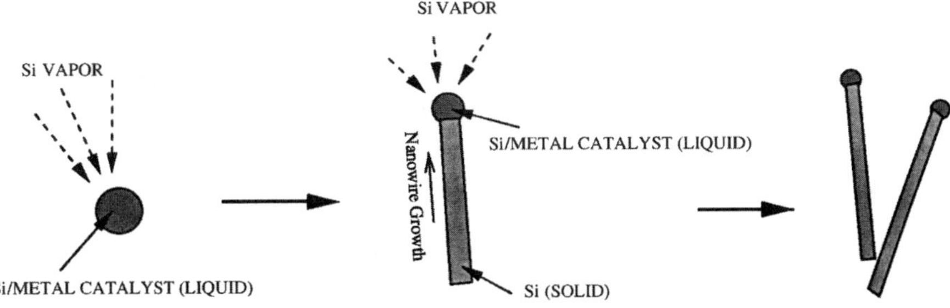

Fig. 10. Schematic illustrating the growth of silicon nanowires by the VLS mechanism.

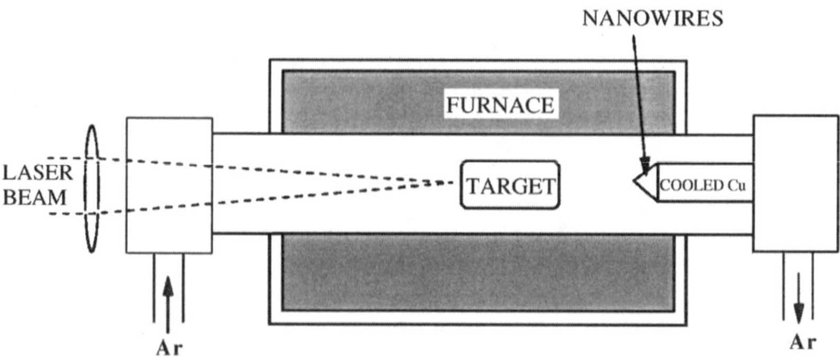

FIG. 11. Schematic of the laser ablation apparatus for the production of nanowires (Morales and Lieber, 1998).

nanowires generated are collected at the cold finger (Morales and Lieber, 1998; Lieber 1998). The growth of nanowires is affected by the reaction temperature and the catalyst material used (Morales and Lieber, 1998; Lieber, 1998; Duan and Lieber, 2000). It is believed that the nanowire growth occurs only when the catalyst cluster remains a liquid (Morales and Lieber, 1998). The wire growth ceases as the catalyst cluster is solidified when it is carried away from the hot zone of the furnace by the gas flow. All the nanowires are found to terminate at one end with catalyst clusters of diameters that are 1.5 to 2 times that of the wires. The nanowires range from 3 nm to tens of nanometers in diameter, with lengths up to tens of micrometers (Morales and Lieber, 1998; Duan and Lieber, 2000). Using the laser ablation technique, Lieber and coworkers have prepared a wide range of semiconducting nanowires, including group IV elements (Si and Ge), III–V compounds (GaAs, GaP, InAs, and InP), II–VI compounds (ZnS, ZnSe, CdS, and CdSe), and binary SiGe alloys (Morales and Lieber, 1998; Duan *et al.*, 2000; Duan and Lieber, 2000). Lee and coworkers also fabricated Si and Ge nanowires using a similar laser ablation apparatus (Zhang *et al.*, 1998b, 2000a).

Figure 12 shows TEM images of silicon nanowires fabricated by Morales and Lieber (1998). The nanowires are produced by laser ablating a $Si_{0.9}Fe_{0.1}$ target. In Fig. 12(a), the darker spheres with a diameter larger than the nanowires are the solidified $FeSi_2$ catalyst clusters that terminate at one end of the nanowires. Electron diffraction pattern indicates that the silicon nanowires are single-crystalline and grow along the [111] direction (Fig. 12(b)) (Morales and Lieber, 1998). The TEM image in Fig. 12(b) shows the silicon nanowire to be uniform in diameter, with a crystalline core that is surrounded by an oxide layer, which may have resulted from the presence of residual oxygen (Morales and Lieber, 1998; Duan *et al.*, 2000). Wang *et al.*

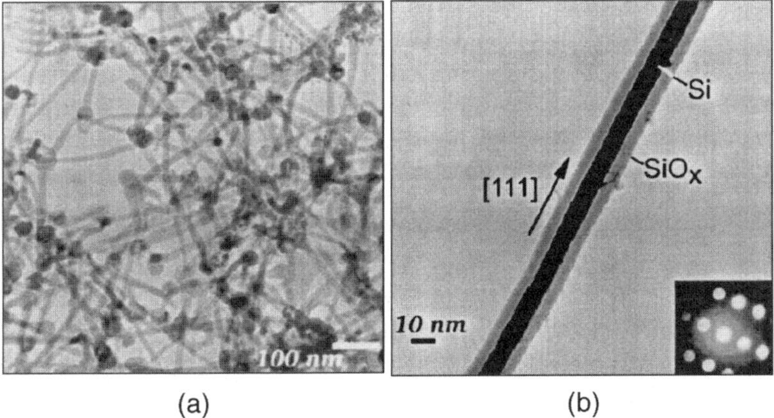

FIG. 12. (a) TEM images of Si nanowires produced after laser ablating a $Si_{0.9}Fe_{0.1}$ target. The dark spheres with a slightly larger diameter than the wires are solidified catalyst clusters (Morales and Lieber, 1998). (b) Diffraction contrast TEM image of a Si nanowire. The crystalline Si core appears darker than the amorphous oxide surface layer. The inset shows the convergent beam electron diffraction pattern recorded perpendicular to the wire axis (Morales and Lieber, 1998).

(1998a,b) found that these oxides played a more important role than metals in assisting the nanowire growth. SiO_x was discovered to be an effective catalyst that significantly increased the yield of silicon nanowires (Zhang *et al.*, 1998b; Wang *et al.*, 1998a, b). Similar yield enhancement was also found in the synthesis of germanium nanowires by laser ablating a germanium powder mixed with GeO_2 (Zhang *et al.*, 2000a). It was observed that targets composed of equal molar ratios of Ge:GeO_2 and Si:SiO_2 in the absence of metal catalysts gave the maximum yields for germanium and silicon nanowires, respectively (Zhang *et al.*, 2000a; Wang *et al.*, 1998a,b). Based on these observations and other TEM studies (Zhang *et al.*, 2000a; Wang *et al.*, 1998a; Lee *et al.*, 1999), an oxide-enhanced nanowire growth mechanism different from the classical VLS mechanism was proposed (Wang *et al.*, 1998a). It was postulated that the nanowire growth was catalyzed by the Ge_mO or Si_mO layer ($m > 1$) on the nanowire tips, which might be in or near their molten states (Lee *et al.*, 1999). Because the oxide-assisted laser ablation method did not require any metal catalysts, the resulting nanowires could achieve a high purity. The germanium and silicon nanowires produced from these catalyst-free targets were noted to grow generally along the [112] crystal direction (Lee *et al.*, 1999).

Although laser ablation methods have been used to fabricate large quantities of single-crystalline semiconducting nanowires with high aspect ratios, the wires obtained are randomly oriented with a variety of morphologies

(Tang *et al.*, 1999) and correspond to broad distributions of wire diameters and lengths. Therefore, it would be important to devote future research efforts toward controlling the wire diameter, length, morphology, and assembly in such synthesis in order to utilize the nanowires derived effectively for potential electronic applications.

C. Other Synthesis Methods

Recently, the VLS growth method has been extended beyond the gas-phase reaction to synthesis of Si nanowires in Si-containing solvent (Holmes *et al.*, 2000). In this case 2.5-nm Au nanocrystals were dispersed in supercritical hexane with a silicon precursor (e.g., diphenylsilane) under a pressure of 200–270 bar at 500°C, at which temperature the diphenylsilane decomposes to Si atoms. The Au nanocrystals serve as seeds for the Si nanowire growth, because they form an alloy with Si, which is in equilibrium with pure Si. It is suggested that the Si atoms would dissolve in the Au crystals until the saturation point is reached; then they are expelled from the particle to form a nanowire with a diameter similar to the catalyst particle. This method has an advantage over the laser-ablated Si nanowire in that the nanowire diameter can be well controlled by the Au particle size, whereas liquid metal droplets produced by the laser ablation process tend to exhibit a much broader size distribution. With this approach, highly crystalline Si nanowires with diameters ranging from 4 nm to 5 nm have been produced by Holmes *et al.* (2000). The crystal orientation of these Si nanowires can be controlled by the reaction pressure.

In addition to the fabrication methods discussed here, special techniques have been developed to prepare a variety of one-dimensional systems for investigating interesting quantum phenomena. For example, quantized conductance (in units of $G_0 = 2e^2/h$, where h is the Planck constant) has been observed through thin metallic nanowires with scanning tunneling microscopy (Pascual *et al.*, 1993, 1995; Brandbyge *et al.*, 1995) and mechanically controlled break junctions (Muller *et al.*, 1992, 1996). These techniques use the same basic principle of pressing two metals together and pulling them out to form a nanometer-sized wire between the two contacts. One-dimensional silicon nanostructures have also been fabricated on a silicon-on-insulator (SOI) substrate by a combination of lithography and orientation-dependent etching (Namatsu *et al.*, 1997). Other approaches include forming nanowires by vapor deposition on cleaved superlattice planes (Arakawa *et al.*, 1996) or by shadow deposition on stepped lattice planes (Sugawara *et al.*, 1997). These fabrication techniques usually require unique experimental configurations or complicated processing and produce nanowires with a very low yield or at a very high cost. Therefore, although the resulting nanostructures may be helpful toward understanding low-dimensional phenomena, their utilization in commercial applications and devices is currently limited.

III. Theoretical Modeling of Nanowire Band Structures

Nanostructured materials exhibit behavior distinct from their bulk counterparts due to quantum confinement effects. As their length scale shrinks to a size comparable to the de Broglie wavelength of electrons, the energy of electrons in the confined direction becomes quantized and forms a discrete energy spectrum. If the separation between the quantized energy levels is much larger than the thermal excitation energy k_BT, virtually all the electrons will occupy the lowest possible states. Because it is not likely to promote electrons to a higher energy state in the absence of an external excitation, the degree of freedom for electrons is quenched in the confined direction, producing a system with reduced dimensionality. The systems confined in one, two, and three dimensions are treated as a two-dimensional (quantum well), a one-dimensional (quantum wire), and a zero-dimensional (quantum dot) electron gas, respectively. The energy quantization changes the band structure of nanostructured materials and alters their optical, magnetic, and electronic properties dramatically. Thus, the effects of quantum confinement on low-dimensional systems are of both fundamental and technological importance. In the following sections, a theoretical framework is presented for the modeling of one-dimensional nanowire systems. The Schrödinger equation has been solved to obtain the band structure of nanowires. Based on the band structure, a semiclassical model is developed to predict various transport properties of nanowire systems.

A. Band Structures of One-Dimensional Systems

The electronic states of nanowire systems exhibit a very different spectrum from that of bulk materials. In order to understand their unique electronic properties, we have modeled the band structure of these one-dimensional systems.

Without loss of generality, we assume a bulk material where the major carriers are electrons with an effective mass $\mathbf{m}_e$. In general, the electron masses are anisotropic, and the effective mass is expressed as a symmetric second-rank tensor. The dispersion relation of the electrons is written as

$$E(\mathbf{k}) = \frac{\hbar^2}{2}\mathbf{k} \cdot \alpha \cdot \mathbf{k}, \tag{2}$$

where $\boldsymbol{k}$ is the wave vector and α is the inverse tensor of $\mathbf{m}_e$. From the effective mass theorem, the envelope wavefunction of electrons, $\psi(\mathbf{r})$, is described by the Schrödinger equation:

$$-\frac{\hbar^2}{2}\nabla \cdot \alpha \cdot \nabla\psi(\mathbf{r}) = E\psi(\mathbf{r}). \tag{3}$$

For nanowires embedded in an insulating matrix with a large band gap (e.g., alumina or mica), electrons are well confined within the wires. Thus, to a good approximation, the electron wavefunction, $\psi(\mathbf{r})$, can be assumed to vanish at the wire boundary.

For an infinitely long wire with a circular cross section of diameter d_w, we take the z axis to be parallel to the wire axis, with the x and y axes lying on the cross-sectional plane. The cylindrical symmetry of the wire is then used to simplify Eq. (3) by making $\alpha_{xy} = \alpha_{yx} = 0$, which can be achieved by a proper rotation about the z axis. The wave function $\psi(\mathbf{r})$ then has the form

$$\psi(\mathbf{r}) = u(x, y)\exp(i\xi \cdot x)\exp(i\eta \cdot y)\exp(ik_z \cdot z), \tag{4}$$

where ξ and η are constants to be determined and k_z is the wave number of the traveling wave in the z direction. By letting $\xi = -(\alpha_{xz}/\alpha_{xx})k_z$ and $\eta = -(\alpha_{yz}/\alpha_{yy})k_z$, Eq. (3) is reduced to a concise second-order differential equation in x and y only:

$$-\frac{\hbar^2}{2}\left(\alpha_{xx}\frac{\partial^2}{\partial x^2} + \alpha_{yy}\frac{\partial^2}{\partial y^2}\right)u(x, y) = \left(E - \frac{\hbar^2 k_z^2}{2m_{zz}}\right)u(x, y), \tag{5}$$

where $m_{zz} = \hat{z} \cdot \mathbf{m_e} \cdot \hat{z}$ is the transport effective mass along the wire axis. Equation (5) is reminiscent of a two-dimensional Schrödinger equation with in-plane effective mass components

$$\begin{aligned} m_x &\equiv \alpha_{xx}^{-1} \\ m_y &\equiv \alpha_{yy}^{-1} \end{aligned} \tag{6}$$

in the x and y directions, respectively. Because $u(x, y)$ must satisfy the boundary condition: $u(x, y) = 0$ when $x^2 + y^2 = (d_w/2)^2$, the eigenvalues of $u(x, y)$ in Eq. (5) are quantized, and the energy of the electrons is written as

$$E_{nm}(k_z) = \varepsilon_{nm} + \frac{\hbar^2 k_z^2}{2m_{zz}}, \tag{7}$$

where ε_{nm} is the eigenvalue of Eq. (5) corresponding to the subband edge eigenstate at $k_z = 0$, labeled by the quantum numbers (n, m).

In a nanowire system, the quantized subband energy ε_{nm} and the transport effective mass m_{zz} along the wire axis are the two most important parameters and determine almost all the electronic properties. Due to the anisotropic carriers and the special geometric configuration (circular wire cross section and high aspect ratio of length to diameter), several approximations were used in earlier calculations to derive ε_{nm} and m_{zz} in bismuth nanowires. In the first calculation carried out by Zhang *et al.* (1998c, 2000b), the quantized energy levels were evaluated by using a cyclotron effective mass approximation for the in-plane effective mass. An improved model was subsequently developed by Sun *et al.* (1999b) based on a square wire cross-section approximation, which allowed an analytical expression to be obtained for the wave functions. However, these two approximations were valid only for less anisotropic carriers, producing significant discrepancies for highly anisotropic systems ($\alpha_{xx}/\alpha_{yy} \gg 1$ or $\alpha_{xx}/\alpha_{yy} \ll 1$). Thus, a numerical approach was developed recently to more accurately determine ε_{nm} and m_{zz} (Lin *et al.*, 2000c).

For a simple case where $\alpha_{xx} = \alpha_{yy}$, the wave function of Eq. (5) has the analytical solution

$$u_{nm}(\mathbf{r}) \sim J_n(\chi_{nm} r) e^{\mathrm{i}n\theta}, \tag{8}$$

where J_n is the nth Bessel function and χ_{nm} is determined by the mth root of $J_n(x \cdot d_w/2) = 0$. The subband energy ε_{nm} corresponding to the wavefunction $u_{nm}(\mathbf{r})$ is given by

$$\varepsilon_{nm} = \frac{\hbar^2}{2} \alpha_{xx} \chi_{nm}^2. \tag{9}$$

For the general case where $\alpha_{xx} \neq \alpha_{yy}$, there are no analytical solutions, and the only possible approach to determine the quantized subband energy ε_{nm} from Eq. (5) is through numerical methods (Lin *et al.*, 2000c). In this instance, a mesh consisting of M concentric circles and N sectors is created within the wire cross section, as shown in Fig. 13. The differential equation of Eq. (5) is then transformed to a set of difference equations based on the grid points on

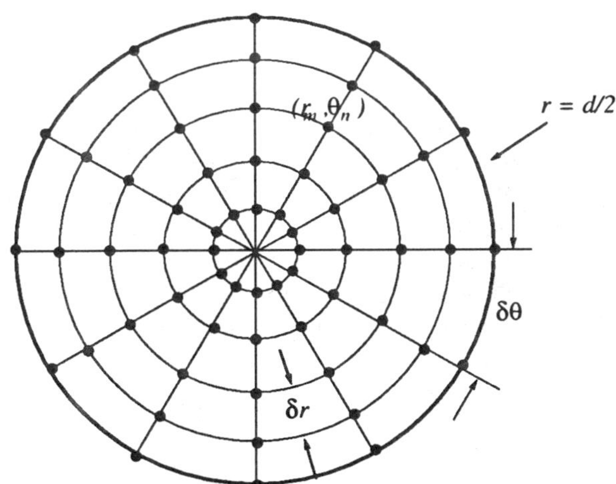

Fig. 13. Schematic of the grid points used to transform the differential equation into a difference equation. The mesh in the circular wire cross section consists of M concentric circles and N sectors. In this figure, $M = 5$ and $N = 12$ (Lin *et al.*, 2000c).

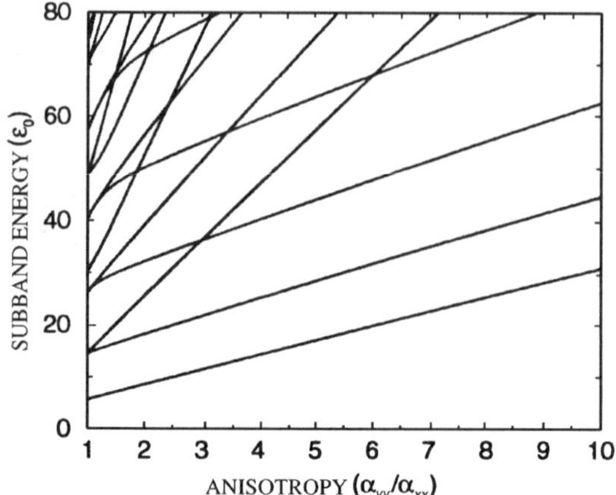

FIG. 14. Calculated subband energies in units of $\varepsilon_0 = 2\alpha_{xx}\hbar^2/d_w^2$ as a function of the in-plane mass anisotropy α_{yy}/α_{xx}. The subband energies of nanowires of various diameters can then be derived from this figure.

the mesh, which can be solved readily with the aid of computers. By refining the mesh, ε_{nm} can be obtained with great accuracy (within 0.1%). Figure 14 shows the calculated subband energies in units of $2\alpha_{xx}\hbar^2/d_w^2$ as a function of the mass anisotropy α_{yy}/α_{xx}.

The density of states (DOS) of electrons in nanowires is derived from Eq. (7) as

$$g(E) = \frac{\sqrt{2m_{zz}}}{\pi\hbar} \sum_{n,m} (E - \varepsilon_{nm})^{-1/2}. \tag{10}$$

Figure 15 shows the calculated DOS for electrons in a 40-nm bismuth nanowire compared to that of bulk bismuth. The DOS in nanowires is a superposition of one-dimensional transport channels, each located at a quantized subband energy ε_{nm}. We note that the DOS in nanowires has sharp peaks at the subband edges, whereas that in a bulk material is a smooth monotonic function of energy. The enhanced DOS at the subband edges of nanowires has important implications for many applications, such as in optics (Black *et al.*, 2000) and thermoelectrics (Hicks and Dresselhaus, 1993).

B. THE SEMIMETAL–SEMICONDUCTOR TRANSITION IN SEMIMETALLIC NANOWIRES

For semimetals in bulk form, such as bismuth, the conduction band overlaps in energy with the valence band, and the electronic properties are

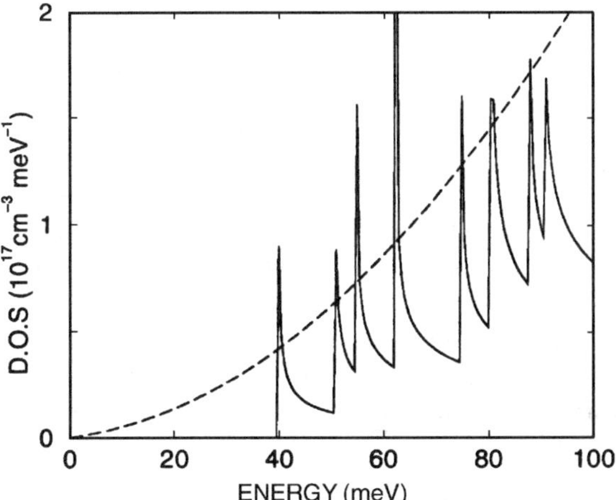

FIG. 15. Calculated effective densities of states for 40-nm bismuth nanowires (solid curve) and bulk bismuth (dashed curve). The zero energy refers to the band edge of bulk bismuth. The nonparabolic effects of the electron carriers are considered in these calculations.

governed by both the electrons and holes. In nanowires, the quantum confinement effects cause the band edge of the electrons to move up in energy (see Fig. 16), whereas the valence band edge decreases in energy. The band-edge energies of electrons and holes shift in opposite directions in bismuth nanowires, decreasing the energy overlap between the conduction band and the valence band (Fig. 16). As the wire diameter continues to decrease, the

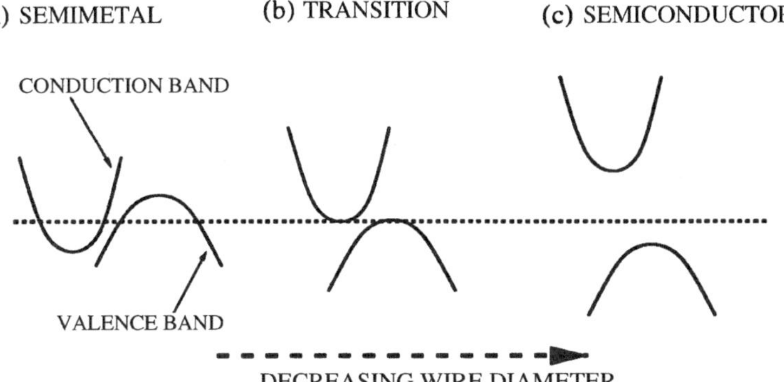

FIG. 16. Schematic illustrating the semimetal–semiconductor transition in nanowires made of semimetals: (a) bulk semimetals with a band overlap between the electrons and the holes, (b) nanowires with the critical wire diameter d_c where the band overlap vanishes, and (c) nanowires with diameters smaller than d_c, exhibiting a bandgap between the conduction and valence bands.

energy overlap eventually vanishes (Fig. 16(b)), producing a band gap between the lowest conduction subband and the highest valence subband. This semimetal-to-semiconductor transition will occur at a critical wire diameter d_c, which depends on the band overlap energy, the electron and hole effective masses, and the crystal orientation along the wire axis for the material of interest. The critical wire diameters for two group V semimetals, bismuth and antimony, are predicted to be about 50 nm and 10 nm (Lin *et al.*, 2000c; Heremans *et al.*, 2001), respectively. This quantum confinement–induced semimetal–semiconductor transition is one of the unique properties of nanowires made of semimetallic materials. Such a transition dramatically alters the electronic properties of nanowires, providing us with new possibilities for manipulating the band structures of materials.

Due to the semimetal–semiconductor transition, the carrier concentration $N(T)$ of nanowires made of semimetallic materials is highly dependent on the wire diameter and temperature and must satisfy the condition $d_w \leq d_c$ to exhibit semiconducting behavior. As an example, Fig. 17 shows the calculated total carrier densities for various bismuth nanowires oriented along the $[01\bar{1}2]$ growth direction as a function of temperature (Lin *et al.*, 2000c). Three different types of temperature dependence for the carrier density are predicted for bismuth nanowires, depending on the wire diameters. For 10-nm bismuth nanowires, which are in the semiconducting regime, the carrier density increases exponentially with temperature. For 80-nm bismuth nanowires, which remain in the semimetallic regime for all temperatures, the carrier density is similar in temperature dependence to bulk bismuth.

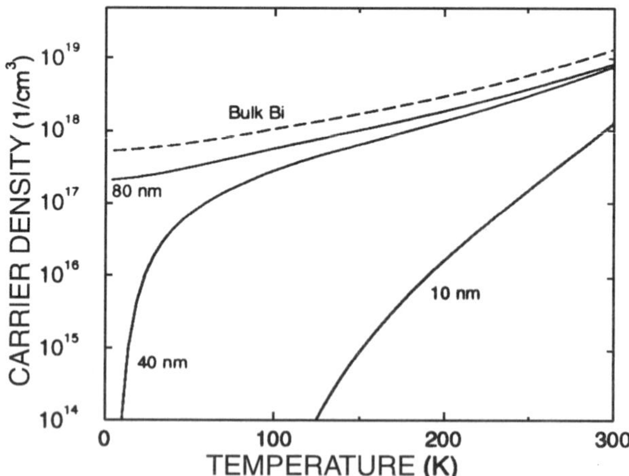

FIG. 17. Calculated total carrier density (electrons and holes) as a function of temperature for bulk 3D bismuth and bismuth nanowires of different diameters oriented along the $[01\bar{1}2]$ direction.

The lower carrier density of the 80-nm nanowires compared to bulk bismuth is due to the smaller band overlap in the former. For the 40-nm bismuth nanowires, the carrier density has a temperature dependence similar to bulk bismuth at high temperatures, but it drops rapidly with decreasing temperature at low temperatures. Because the carrier density is highly dependent on wire diameter, the transport properties of bismuth nanowires are expected to be highly sensitive to wire diameter, as will be shown experimentally in the section "temperature-dependent resistivity of nanowires."

IV. Transport Properties

The transport phenomena in low-dimensional systems can be roughly divided into two categories: ballistic transport and diffusive transport. Ballistic transport occurs when the electrons travel across a nanowire without any scattering. In this case, the conduction is determined mainly by the contacts between the nanowire and the external circuit, and the conductance is quantized into a universal conductance unit, $G_0 = 2e^2/h$ (Wharam *et al.,* 1988; van Wees *et al.,* 1988). Ballistic transport is usually observed in very short quantum wires, such as those produced by mechanically controlled break junctions (Muller *et al.,* 1992, 1996) or by scanning tunneling microscopy (Pascual *et al.,* 1993, 1995; Brandbyge *et al.,* 1995), whereby the wire length is much shorter than the electron mean free path and the conduction is a pure quantum phenomenon. An additional requirement for ballistic transport is that $k_B T \ll \varepsilon_j - \varepsilon_{j-1}$, where $\varepsilon_j - \varepsilon_{j-1}$ is the subband separation between the j and $j-1$ subband energy levels. On the other hand, for nanowires with lengths much longer than the carrier mean free path, the electrons or holes undergo numerous scattering events when they travel along the wire. In this case, electron transport is in the diffusive regime, and conduction is dominated by scattering due to phonons (lattice vibrations), boundary or lattice defects, and impurity atoms.

The ballistic transport of one-dimensional systems has been extensively studied since the discovery of quantized conductance (Wharam *et al.,* 1988; van Wees *et al.,* 1988). In contrast, due to the difficulty in fabricating long nanowires ($> 1\ \mu$m) and the inadequacy of conventional experimental techniques, there have been fewer experimental and theoretical studies on transport phenomena in the diffusive regime. Recently, a semiclassical transport model based on the band structure of nanowires was developed for systems in which scattering events are not negligible (Lin *et al.,* 2000b,c). The diffusive transport model has been applied to bismuth and antimony nanowire systems (Lin *et al.,* 2000b; Heremans *et al.,* 2001) and agrees well with the experimental findings.

A. Semiclassical Model

The diffusive transport phenomena in nanowires can be described by a semiclassical model based on the Boltzmann transport equation. For carriers in a one-dimensional subband, important transport coefficients, such as the electrical conductivity, σ, the Seebeck coefficient, S, and the thermal conductivity, κ_e, are derived as (Sun *et al.*, 1999b; Ashcroft and Mermin, 1976a)

$$\sigma = L^{(0)} \tag{11}$$

$$S = -\frac{1}{eT}\frac{L^{(1)}}{L^{(0)}} \tag{12}$$

$$\kappa_e = \frac{1}{e^2T}\left(L^{(2)} - \frac{(L^{(1)})^2}{L^{(0)}}\right), \tag{13}$$

where T is temperature and

$$L^{(\alpha)} = e^2 \int \frac{4\,dk}{\pi^2 d_w^2}\left(-\frac{df}{dE}\right)\tau(k)v(k)v(k)(E(k) - E_f)^\alpha, \tag{14}$$

where $\alpha = 0, 1, 2$, k is the wave vector along the transport direction, $E(k)$ denotes the carrier dispersion relation, $v(k)$ is the group velocity, $\tau(k)$ is the relaxation time, E_f is the Fermi energy, and $f(E)$ is the Fermi–Dirac distribution function. Assuming a parabolic dispersion relation, the transport elements $L^{(\alpha)}$ in Eqs. (11)–(13) are derived as

$$L^{(0)} = D\left[\frac{1}{2}F_{-1/2}\right] \tag{15}$$

$$L^{(1)} = \begin{cases} (k_BT)D\left[\frac{3}{2}F_{1/2} - \frac{1}{2}\varsigma^* F_{-1/2}\right] & \text{(for electrons)} \\ -(k_BT)D\left[\frac{3}{2}F_{1/2} - \frac{1}{2}\varsigma^* F_{-1/2}\right] & \text{(for holes)} \end{cases} \tag{16}$$

$$L^{(2)} = (k_BT)^2 D\left[\frac{5}{2}F_{3/2} - 3\varsigma^* F_{1/2} + \frac{1}{2}\varsigma^{*2}F_{-1/2}\right], \tag{17}$$

where D is given by

$$D = \frac{8e}{\pi^2 d_w^2} = \left(\frac{2m^* k_B T}{\hbar^2}\right)^{1/2}\mu, \tag{18}$$

in which μ is the carrier mobility along the nanowire and

$$F_j = \int_0^\infty \frac{x^j\,dx}{\exp(x - \varsigma^*) + 1} \tag{19}$$

denotes the Fermi–Dirac related functions, with fractional indices. $j = -\frac{1}{2}, \frac{1}{2}, \frac{3}{2}$. The reduced chemical potential, ς^*, is defined as

$$\varsigma^* = \begin{cases} \left(E_f - \varepsilon_e^{(0)}\right)/k_B T & \text{(for electrons)} \\ \left(\varepsilon_h^{(0)} - E_f\right)/k_B T & \text{(for holes),} \end{cases} \tag{20}$$

where $\varepsilon_e^{(0)}$ and $\varepsilon_h^{(0)}$ denote the band edges for electrons and holes, respectively. In considering the transport properties of real one-dimensional nanowires, contributions from all the subbands near the Fermi energy should be included, and the $L^{(\alpha)}$s in Eqs. (11)–(13) should be replaced by the sum of contributions from each subband i, $L_{total}^{(\alpha)} = \sum_i L_i^{(\alpha)}$, to obtain the various transport coefficients.

It should be noted that the carrier mobility in nanowires is lower than that in bulk single-crystalline material due to possible scattering at wire and grain boundaries, uncontrolled impurities, and lattice defects. The overall effect of this additional scattering is taken into account by Matthiessen's rule (Ashcroft and Mermin, 1976b),

$$\frac{1}{\mu_{\text{tot}}(T)} = \frac{1}{\mu_{\text{bulk}}(T)} + \frac{1}{\mu_{\text{bound}}} + \frac{1}{\mu_{\text{imp}}(T)}, \tag{21}$$

where μ_{bulk} is the carrier mobility in bulk crystalline material and the terms μ_{bound}^{-1} and μ_{imp}^{-1} account for boundary scattering and charged impurity scattering, respectively. Also, μ_{bound}^{-1} is usually assumed to be independent of temperature, whereas μ_{imp}^{-1} has a temperature dependence of $T^{1.5}$ for most charged impurity scattering processes.

B. Temperature-Dependent Resistivity of Nanowires

A number of transport measurements have been performed on nanowires prepared by the template-assisted approach. The nanowire-embedded template provides a convenient package for making electrical contacts on both ends of the wires, so that two-point transport measurements can be performed. Transport properties of various nanowire arrays have been measured by Zhang *et al.* (1998c, 2000b), Heremans *et al.* (1998, 2000), Liu *et al.* (1998b), Huber *et al.* (1999), Hong *et al.* (1999), Lin *et al.* (2000b), and Sun *et al.* (2000). Although a two-point resistance measurement has the advantage of simplicity, an absolute resistivity value for the nanowires cannot be determined by this approach, because the number of wires in the template contributing to the conduction measurement is not known. Progress was recently made to characterize the absolute resistivity of a single nanowire

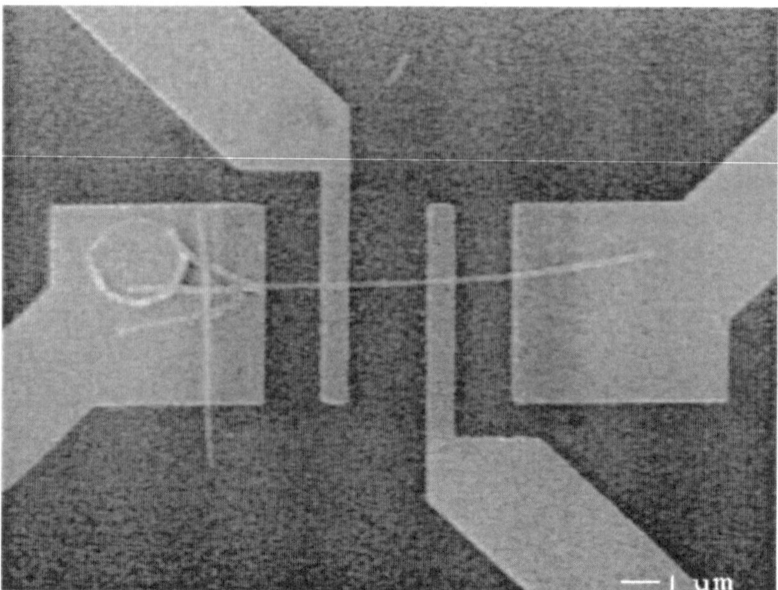

FIG. 18. SEM image of a 70-nm bismuth nanowire with four electrodes attached to the nanowire. The circle on the large left electrode is a reference point used to find the nanowire and to attach electrodes to it by a lithographic process (Cronin *et al.*, 1999).

via a four-point setup (Cronin *et al.*, 1999, 2000), which could provide more physical information on the properties of nanowires than the normalized resistance of a nanowire array. Figure 18 shows an SEM image of a four-point electrode patterned on a 70-nm Bi nanowire (Cronin *et al.*, 1999). The circular dot in Fig. 18 represents one of the prepatterned grid points used to locate the nanowires on the substrate. In this case, nanoelectrodes were patterned by electron-beam lithography on top of a single nanowire on a substrate coated with a thin insulating layer (Cronin *et al.*, 1999, 2000). The electrodes consisted of a gold layer ($\sim$ 1000 Å thick) and a thin adhesive layer, and their processing followed a standard *lift-off* method. We note that most nanowires would undergo surface oxidation upon removal from the template. The surface oxide layer on the nanowires imposes a serious problem in making nanoelectrical contacts; efforts are currently being devoted to tackling this challenge.

Figure 19(a) shows the temperature dependence of resistance $R(T)$ for bismuth nanowire arrays ($d_w = 7 - 200$ nm) synthesized by vapor deposition and measured by Heremans *et al.* (2000). Hong *et al.* (1999) reported similar resistance measurements on bismuth wires of larger diameters (200 nm to 2 μm) prepared by electrochemical deposition (Fig. 19(b)). These two studies

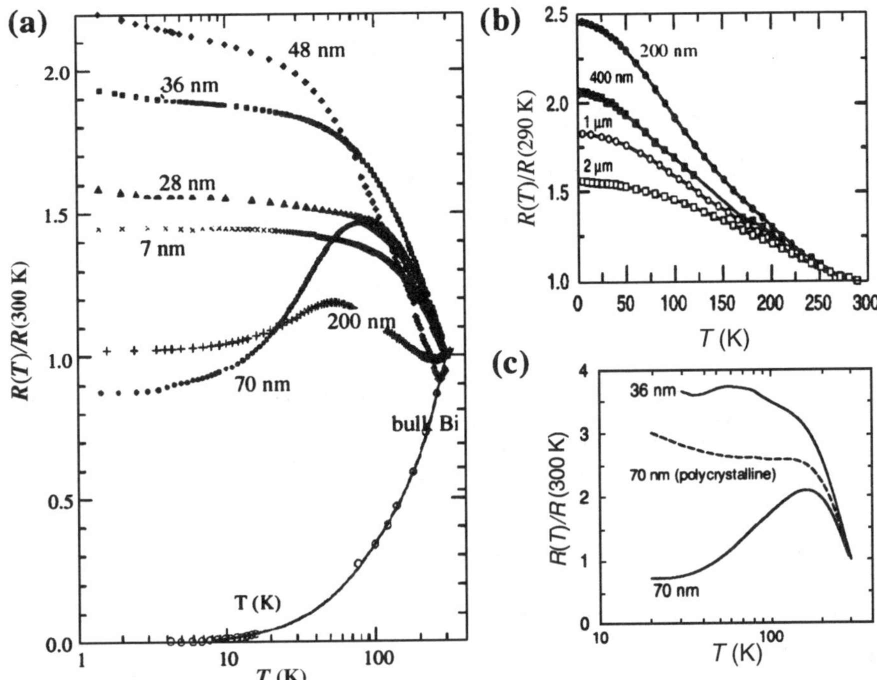

FIG. 19. (a) Measured temperature dependence of resistance for bismuth nanowire arrays of various wire diameters d_w (Heremans *et al.*, 2000). (b) $R(T)/R(290\ K)$ for bismuth wires of larger d_w measured by Hong *et al.* (1999). (c) Calculated $R(T)/R(300\ K)$ of 36-nm and 70-nm bismuth nanowires (Lin *et al.*, 2000b). The dashed curve refers to a 70-nm polycrystalline wire with increased boundary scattering.

showed that $R(T)$ of nanowires is highly sensitive to d_w and is very different from that of bulk bismuth. In Fig. 19(a), the temperature dependence of resistance for bismuth nanowires exhibits dissimilar trends for $d_w > 50$ nm and $d_w \leq 50$ nm. Based on the semiclassical transport model and the band-structure model of bismuth nanowires, $R(T)/R(300\text{ K})$ was calculated for 36-nm and 70-nm wires. The two wire diameters were chosen to represent semiconducting and semimetallic bismuth nanowires, respectively. The solid curves in Fig. 19(c) (Lin *et al.*, 2000b) illustrate that the calculated $R(T)/R(300\text{ K})$ trends are consistent with those obtained experimentally (Fig. 19(a)). We note that the nonmonotonic $R(T)$ behavior for semimetallic nanowires observed by Heremans *et al.*, 2000 in Fig. 19(a) is not found in the wires of Fig. 19(b). This inconsistency may be due to the crystal quality differences in the bismuth wires prepared by the two different approaches, which can be accounted for by μ_{bound} in the transport model. Instead of the nonmonotonic behavior exhibited by single-crystalline semimetallic nanowires in Fig. 19(a),

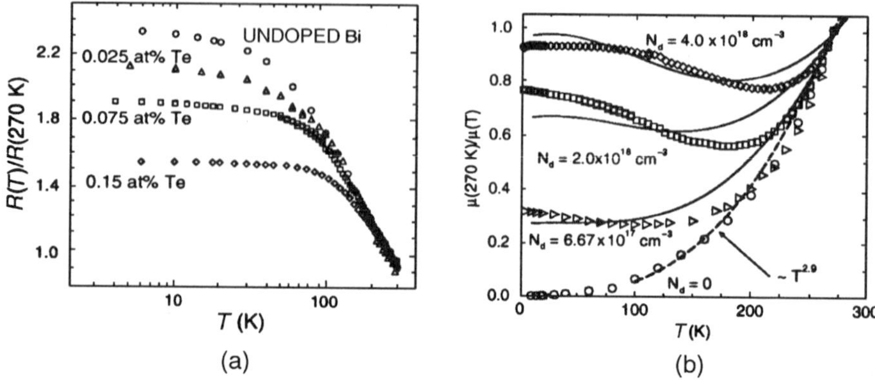

FIG. 20. (a) Measured $R(T)/R(270\,K)$ for 40-nm bismuth nanowires prepared with alloys of different Te doping levels. (b) The calculated temperature dependence of μ_{avg}^{-1} for 40-nm undoped and Te-doped bismuth nanowires of different N_d. The dashed and solid lines are fitting curves corresponding to undoped and Te-doped Bi nanowires, respectively.

$R(T)$ is predicted to display a monotonic temperature dependence at a high defect level. This is illustrated by the dashed curve in Fig. 19(c) for polycrystalline 70-nm bismuth wires. Because the nanowires prepared by electrochemical deposition were found to be polycrystalline, their carriers would experience more boundary scattering, resulting in the monotonic $R(T)$ behavior noted experimentally in Fig. 19(b).

The same transport model has also been extended to describe the properties of Te-doped bismuth nanowires and antimony nanowires. Te, a group VI element, is an electron donor in bismuth. Figure 20(a) shows the measured $R(T)/R(270\text{ K})$ for 40-nm Bi nanowires with the Te concentrations used to form the Bi–Te alloys for nanowire synthesis. The actual Te concentrations in the bismuth nanowires would be smaller than these nominal values because some Te atoms would segregate to the wire boundary during alloy solidification (Zhang *et al.*, 2000b). For simplicity, we assume that about 10% of the Te dopants in the alloy melt are present in the final nanowire product and that each Te atom donates one electron to the conduction band of bismuth. Therefore, 0.025, 0.075, and 0.15 at% Te-doped Bi alloys give rise to donor concentrations N_d of 6.67×10^{17}, 2.0×10^{18}, and 4.0×10^{18} cm^{-3} in the respective nanowires. Based on the measured $R(T)$ in Fig. 20(a) and the calculated temperature-dependent carrier density, the average carrier mobility, $\mu_{\text{avg}}(T)$, of the Te-doped bismuth nanowires can be obtained. Analogous to Eq. (21), μ_{avg}^{-1} for doped bismuth nanowires can be related to the various scattering processes by $\mu_{\text{doped}}^{-1}(T) = \mu_{\text{undoped}}^{-1}(T) + \mu_{\text{imp}}^{-1}(T) + \mu_{\text{defect}}^{-1}$, where μ_{undoped} is the average mobility of the undoped bismuth nanowires with the same diameter and μ_{imp}^{-1} and μ_{defect}^{-1} are associated with the increased ionized impurity scattering and the higher defect level in Te-doped

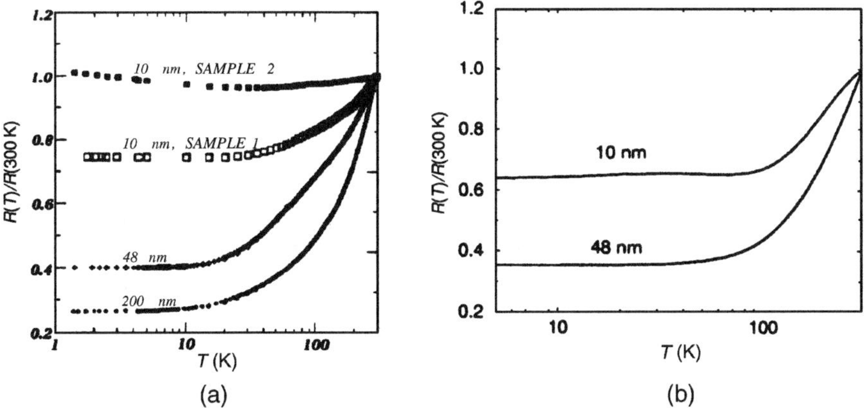

FIG. 21. (a) Temperature dependence of the resistance measured for various antimony nanowires, normalized to the resistance at 300 K. (b) Calculated $R(T)/R(300\,K)$ for 10-nm and 48-nm antimony nanowires.

bismuth nanowires, respectively. The $\mu_{avg}{}^{-1}(T)$ values calculated are shown in Fig. 20(b) as solid curves, which are qualitatively consistent with the experimental results (denoted by symbols).

Like bismuth, antimony is semimetallic in bulk form. The band overlap between electrons and holes in antimony is about 180 meV at 4 K, which is about five times larger than that in bismuth ($\sim$ 38 meV). Therefore, the semimetal–semiconductor transition would occur at a smaller wire diameter in antimony nanowires ($\sim$ 10 nm) compared to bismuth nanowires ($\sim$ 50 nm). Figure 21(a) shows the temperature dependence of the resistance for antimony nanowires prepared by vapor deposition (Heremans *et al.,* 2001). One of the 10-nm nanowire arrays has less resistance variation with temperature than the other, probably due to differences in impurity content or wire diameter distribution within the anodic alumina templates. Figure 21(b) illustrates the modeled $R(T)$ curves for 10-nm and 48-nm antimony nanowire arrays, which display trends that are qualitatively consistent with the experimental results shown in Fig. 21(a) (Heremans *et al.,* 2001). The fact that the measured $R(T)$ of antimony and bismuth nanowires can be explained by the same transport model suggests that the different temperature dependences of the resistance between nanowires and bulk materials arise from both quantum finite-size effects and classical finite-size effects. The classical finite-size effect decreases the carrier mobility by limiting the carrier mean free path, whereas the quantum confinement effect alters the band structure (especially for semimetals) and significantly changes the carrier density. Together, these two factors determine the temperature dependence of resistance in the nanowire systems. Finite-size effects in nanowires have also been observed in the magnetoresistance measurements of bismuth and nickel nanowire arrays (Fasol, 1998; Sun *et al.*, 2000).

V. Summary

Recent developments in the fabrication of nanowires have led to significant advances in research on these low-dimensional systems. The laser-assisted synthesis has successfully produced a variety of semiconducting nanowires with ultrafine diameters. In contrast, most metallic nanowires have been prepared in a regular array with a well-controlled wire-packing geometry via the template-assisted approach. Nanoporous templates provide a useful matrix for handling nanowires and for device integration. The semimetallic nanowires of bismuth and antimony have been shown to undergo a transition to semiconducting behavior with decreasing wire diameters. A generalized theoretical framework has been developed to predict the band structure and transport properties of nanowire systems. Theoretical calculations indicate that nanowires exhibit a very different band structure from that of their bulk counterparts, resulting in unusual optical, electronic and thermoelectric properties. Transport studies suggest that quantum confinement effects significantly perturb the carrier density, whereas the carrier mean free path is limited mainly by the wire diameter. Consequently, transport phenomena in nanowire systems can be manipulated by tuning the diameter of these one-dimensional structures.

Nanowire systems present a great challenge for the understanding and utilization of low-dimensional materials. Although device fabrication based on nanowire systems is still in its infancy, these nanostructures have demonstrated a significant potential for technological breakthroughs with their unique band structure and transport properties. They offer tremendous research opportunities at the frontiers of nanotechnology. Future advances in this field would entail further interdisciplinary efforts toward developing novel synthesis method and control of self-assembled nanostructures, detailed understanding and theoretical modeling of quantum confinement effects, and sophisticated characterization of nanostructures with high aspect ratios.

ACKNOWLEDGMENTS

The authors acknowledge helpful discussions with Dr. Z. Zhang, Dr. G. Dresselhaus, Dr. J. Heremans, O. Rabin, M. R. Black, and S. B. Cronin. They are grateful to the NSF, the U.S. Navy, the MURI program, and DARPA for financial support.

REFERENCES

Adamson, A. W., "Physical Chemistry of Surfaces." Wiley, New York, 1982, p. 338.

AlMawlawi, D., Coombs, N., and Moskovits, M., Magnetic properties of Fe deposited into anodic aluminum-oxide pores as a function of particle size. *J. Appl. Phys.* **70,** 4421 (1991).

Arakawa, T., Watabe, H., Nagamune, Y., and Arakawa, Y., Fabrication and microscopic photoluminescence imaging of ridge-type InGaAs quantum wires grown on a (110) cleaved plane of AlGaAs/GaAs superlattice. *Appl. Phys. Lett.* **69,** 1294 (1996).

Ashcroft, N. W., and Mermin, N. D., "Solid State Physics." Holt, Rinehart and Winston, New York, 1976a, Chap. 13.

Ashcroft, N. W., and Mermin, N. D., "Solid State Physics." Holt, Rinehart and Winston, New York, 1976b, Chap. 16.

Beck, J. S., Vartuli, J. C., Roth, W. J., Leonowicz, M. E., Kresge, C. T., Schmitt, K. D., Chu, C. T-W., Olson, D. H., Sheppard, E. W., McCullen, S. B., Higgins, J. B., and Schlenker, J. L., A new family of mesoporous molecular sieves prepared with liquid crystal templates. *J. Am. Chem. Soc.* **114,** 10834 (1992).

Berry, A. D., Tonucci, R. J., and Fatemi, M., Fabrication of GaAs and InAs wires in nanochannel glass. *Appl. Phys. Lett.* **69,** 2846 (1996).

Bhattacharrya, S., and Saha, S. K., Nanowire formation in a polymeric film. *Appl. Phys. Lett.* **76,** 3896 (2000).

Black, M. R., Lin, Y.-M., Dresselhaus, M. S., Tachibama, M., Fang, S., Rabin, O., Ragot, F., Eklund, P. C., and Dunn, B., Measuring the dielectric properties of nanostructures using optical reflection and transmission: bismuth nanowires in porous alumina. *MRS Symp. Proc.* **581,** 623 (2000).

Blondel, A., Meier, J. P., Doudin, B., and Ansermet, J.-Ph., Giant magnetoresistance of nanowires of multilayers. *Appl. Phys. Lett.* **65,** 3019 (1994).

Brandbyge, M., Schiøtz, J., Sörensen, M. R., Stoltze, P., Jacobsen, K. W., Nørskov, J. K., Olesen, L., Laegsgaard, E., Stensgaard, I., and Besenbacher, F., Quantized conductance in atom-sized wires between two metals. *Phys. Rev. B* **52,** 8499 (1995).

Braun, E., Eichen, Y., Sivan, U., and Ben-Yoseph, G., DNA-templated assembly and electrode attachment of a conducting silver wire. *Nature* **391,** 775 (1998).

Cheng, G. S., Zhang, L. D., Zhu, Y., Fei, G. T., Li, L., Mo, C. M., and Mao, Y. Q., Ordered nanostructure of single-crystalline GaN nanowires in a honeycomb structure of anodic alumina. *Appl. Phys. Lett.* **75,** 2455 (1999).

Cheng, G. S., Zhang, L. D., Chen, S. H., Li, Y., Li, L., Zhu, X. G., Zhu, Y., Fei, G. T., and Mao, Y. Q., Ordered nanostructure of single-crystalline GaN nanowires in a honeycomb structure of anodic alumina. *J. Mater. Res.* **15,** 347 (2000).

Chou, S. Y., Krauss, P. R., and Kong, L. S., Nanolithograpically defined magnetic structures and quantum magnetic disk. *J. Appl. Phys.* **79,** 6101 (1996).

Cronin, S. B., Lin, Y.-M., Koga, T., Sun, X., Ying, J. Y., and Dresselhaus, M. S., Thermoelectric investigation of bismuth nanowires, *in* "The 18th International Conference on Thermoelectrics: ICT Symposium Proceedings" (G. Chen, Ed.), p. 554. IEEE, Piscataway, NJ, 1999.

Cronin, S. B., Lin, Y.-M., Koga, T., Ying, J. Y., and Dresselhaus, M. S., Transport measurements of individual bismuth nanowires. *MRS Symp. Proc.* **582,** 10.4 (2000).

Davydov, D. N., Sattari, P. A., AlMawlawi, D., Osika, A., and Haslett, T. L., Field emitters based on porous aluminum oxide templates. *J. Appl. Phys.* **86,** 3983 (1999).

Diggle, J. W., Downie, T. C., and Goulding, C. W., Anodic oxide films on aluminum. *Chem. Rev.* **69,** 365 (1969).

Duan, X., and Lieber, C. M., General synthesis of compound semiconductor nanowires. *Adv. Mater.* **12,** 298 (2000).

Duan, X., Wang, J., and Lieber, C. M., Synthesis and optical properties of gallium arsenide nanowires. *Appl. Phys. Lett.* **76,** 1116 (2000).

Fasol, G., Nanowires: Small is beautiful. *Science* **280,** 545 (1998).

Ferain, E., and Legras, R., Track-etched membrane-dynamics of pore formation. *Nucl. Instrum. Methods B* **84,** 539 (1993).

Ferre, R., Ounadjela, K., George, J. M., Piraux, L., and Dubois, S., Magnetization processes in nickel and cobalt electrodeposited nanowires. *Phys. Rev. B* **56,** 14066 (1997).

Foss, C. A., Jr., Tierney, M. J., and Martin, C. R., Template synthesis of infrared-transparent metal microcylinders—comparison of optical properties with the predictions of effective medium theory. *J. Phys. Chem.* **96,** 9001 (1992).

Govyadinov, A. N., and Zakhvitcevich, S. A., Field emitter arrays based on natural self-organized porous anodic alumina. *J. Vac. Sci. Technol. B* **16,** 1222 (1998).

Han, Y.-J., Kim, J. M., and Stucky, G. D., Preparation of noble metal nanowires using hexagonal mesoporous silica SBA-15. *Chem. Mater.* **12,** 2068 (2000).

Heremans, J., Thrush, C. M., Zhang, Z., Sun, X., Dresselhaus, M. S., Ying, J. Y., and Morelli, D. T., Magnetoresistance of bismuth nanowire arrays: a possible transition from 1D to 3D localization. *Phys. Rev. B* **58,** R10091 (1998).

Heremans, J., Thrush, C. M., Lin, Y.-M., Cronin, S. B., Zhang, Z., Dresselhaus, M. S., and Mansfield, J. F., Bismuth nanowire arrays: Synthesis and galvanomagnetic properties. *Phys. Rev. B* **61,** 2921 (2000).

Heremans, J., Thrush, C. M., Lin, Y.-M., Cronin, S., and Dresselhaus, M. S., Transport properties of antimony nanowires. *Phys. Rev. B.* **63,** 5406 (2001).

Hicks, L. D., and Dresselhaus, M. S., Thermoelectric figure of merit of a one-dimensional conductor. *Phys. Rev. B* **47,** 16631 (1993).

Holmes, J. D., Johnston, K. P., Doty, R. C., and Korgel, B. A., Control of thickness and orientation of solution-grown silicon nanowires. *Science* **287,** 1471 (2000).

Hong, K., Yang, F. Y., Liu, K., Reich, D. H., Searson, P. C., and Chien, C. L., Giant positive magnetoresistance of Bi nano wire arrays in high magnetic fields. *J. Appl. Phys.* **85,** 6184 (1999).

Hornyak, G. L., Patrissi, C. J., and Martin, C. R., Fabrication, characterization, and optical properties of gold nanoparticle/porous alumina composites: the nonscattering Maxwell-Garnett. *J. Phys. Chem.* **101,** 1548 (1997).

Huber, C. A., Huber, T. E., Sadoqi, M., Lubin, J. A., Manalis, S., and C. B., Prater, Nanowire array composites. *Science* **263,** 800 (1994).

Huber, T. E., Graf, M. J., and Foss, C. A., Jr., Low contact resistance 30 nm and 200 nm diameter Bi wire array composites, *in* "The 18th International Conference on Thermoelectrics: ICT Symposium Proceedings" (G. Chen, Ed.), p. 558. IEEE, Piscataway, NJ, 1999.

Huber, T. E., Graf, M. J., Foss, C. A., Jr., and Constant, P., Processing and characterization of high-conductance bismuth wire array composites. *J. Mater. Res.* **15,** 1816 (2000).

Iwasaki, T., Motoi, T., and Den, T., Mutiwalled carbon nanotubes growth in anodic alumina nanoholes. *Appl. Phys. Lett.* **75,** 2044 (1999).

Jessensky, O., Müller, F., and Gösele, U., Self-organized formation of hexagonal pore arrays in anodic alumina. *Appl. Phys. Lett.* **72,** 1173 (1998).

Keller, F., Hunter, M. S., and Robinson, D. L., Structural features of oxide coating on aluminum. *J. Electrochem. Soc.* **100,** 411 (1953).

Kouklin, N., Bandyopadhyay, S., Tereshin, S., Varfolomeev, A., and Zaretsky, D., Electronic bistability in electrochemically self-assembled quantum dots: A potential nonvolatile random access memory. *Appl. Phys. Lett.* **76,** 460 (2000).

Lee, S. T., Zhang, Y. F., Wang, N., Tang, Y. H., Bello, I., Lee, C. S., and Chung, Y. W., Semiconductor nanowires from oxides. *J. Mater. Res.* **14,** 4503 (1999).

Li, A. P., Müller, F., Birner, A., Neilsch, K., and Gösele, U., Hexagonal pore arrays with a 50–420 nm interpore distance formed by self-organization in anodic alumina. *J. Appl. Phys.* **84,** 6023 (1998a).

Li, F., Zhang, L., and Metzger, R. M., On the growth of highly ordered pores in anodized aluminum oxide. *Chem. Mater.* **10,** 2470 (1998b).

Li, J., Papadopoulos, C., Xu, J. M., and Moskovits, M., Highly ordered carbon nanotube arrays for electronic applications. *Appl. Phys. Lett.* **75,** 367 (1999).

Li, Y., Holland, E. R., and Wilshaw, P. R., Synthesis of high density arrays of nanoscaled gridded field emitters based on anodic alumina. *J. Vac. Sci. Technol. B* **18,** 994 (2000).

Lieber, C. M., One-dimensional nanostructures: chemistry, physics and applications. *Solid State Commun.* **107,** 607 (1998).

Lin, Y.-M., Sun, X., Cronin, S. B., Zhang, Z., Ying, J. Y., and Dresselhaus, M. S., Fabrication and transport properties of Te-doped bismuth nanowire arrays. *MRS Symp. Proc.* **582,** 10.3 (2000a).

Lin, Y.-M., Cronin, S. B., Ying, J. Y., Dresselhaus, M. S., and Heremans, J. P., Transport properties of Bi nanowire arrays. *Appl. Phys. Lett.* **76,** 3944 (2000b).

Lin, Y.-M., Sun, X., and Dresselhaus, M. S., Investigation of thermoelectric transport properties of cylindrical Bi nanowires. *Phys. Rev. B* **62,** 4610 (2000c).

Liu, K., Chien, C. L., Searson, P. C., and Kui, Y. Z., Structural and magneto-transport properties of electrodeposited bismuth nanowires. *Appl. Phys. Lett.* **73,** 1436 (1998a).

Liu, K., Chien, C. L., and Searson, P. C., Finite-size effects in bismuth nanowires. *Phys. Rev. B* **58,** 14681 (1998b).

Martin, C. R., Nanomaterials: A membrane-based synthetic approach. *Science* **266,** 1961 (1994).

Masuda, H., Yamada, H., Satoh, M., Asoh, H., Nakao, M., and Tamamura, T., Highly ordered nanochannel-array architecture in anodic alumina. *Appl. Phys. Lett.* **71,** 2772 (1997).

Masuda, H., Ohya, M., Asoh, H., and Nakao, M., Photonic crystal using anodic porous alumina. *Jpn. J. Appl. Phys.* **38,** 1403 (1999).

Morales, A. M., and Lieber, C. M., A laser ablation method for the synthesis of crystalline semiconductor nanowires. *Science* **279,** 208 (1998).

Muller, C. J., van Ruitenbeek, J. M., and de Jongh, L. J., Conductance and supercurrent discontinuities in atomic-scale metallic constrictions of variable width. *Phys. Rev. Lett.* **69,** 140 (1992).

Muller, C. J., Krans, J. M., Todorov, T. N., and Reed, M. A., Quantization effects in the conductance of metallic contacts at room temperature. *Phys. Rev. B* **53,** 1022 (1996).

Namatsu, H., Horiguchi, S., Nagase, M., and Kurihara, K., Fabrication of one-dimensional nanowire structures utilizing crystallographic orientation in silicon and their conductance characteristics. *J. Vac. Sci. Technol. B* **15,** 1688 (1997).

Nötzel, R., Ledentsov, N. N., Daweeritz, L., and Ploog, K., Semiconductor quantum-wire structures directly grown on high-index surfaces. *Phys. Rev. B* **45,** 3507 (1992).

Omi, H., and Ogino, T., Self-assembled Ge nanowires grown on Si(113). *Appl. Phys. Lett.* **71,** 2163 (1997).

O'Sullivan, J. P., and Wood, G. C., The morphology and mechanism of formation of porous anodic films on aluminum. *Proc. Royal Soc. London A* **317,** 511 (1970).

Ozin, G. A., Nanochemistry: synthesis in diminishing dimensions. *Adv. Mater.* **4,** 612 (1992).

Pascual, J. I., Mendez, J., Gómez-Herrero, J., Baró, A. M., Garcia, N., and Binh, V. T., Quantum contact in gold nanostructures by scanning tunneling microscopy. *Phys. Rev. Lett.* **71,** 1852 (1993).

Pascual, J. I., Mendez, J., Gómez-Herrero, J., Baró, A. M., Garcia, N., Landman, U., Luedtke, W. D., Bogachek, E. N., and Cheng, H. P., Properties of metallic nanowires—from conductance quantization to localization. *Science* **267,** 1793 (1995).

Peng, Y., Zhang, H.-L., Pan, S.-L, and Li, H.-L., Magnetic properties and magnetization reversal of α-Fe nanowires. *J. Appl. Phys.* **87,** 7405 (2000).

Piraux, L., George, I. M., Despres, J. F., Leroy, C., Ferain, E., Legras, R., Ounadjela, K., and Fert, A., Giant magnetoresistance in magnetic multilayered nanowires. *Appl. Phys. Lett.* **65,** 2484 (1994).

Piraux, L., Dubois, S., Duvail, J. L., and Radulescu, A., Fabrication and properties of organic and metal nanocylinders in nanoporous membranes. *J. Mater. Res.* **14,** 3042 (1999).

Routkevitch, D., Bigioni, T., Moskovits, M., and Xu, J. M., Electrochemical fabrication of CdS nanowire arrays in porous anodic aluminum oxide templates. *J. Phys. Chem.* **100,** 14307 (1996a).

Routkevitch, D., Tager, A. A., Haruyama, J., AlMawlawi, D., Moskovits, M., and Xu, J. M., Nonlithographic nanowire arrays: Fabrication, physics, and device applications. *IEEE Trans. Electron. Dev.* **43,** 1646 (1996b).

Sugawara, A., Coyle, T., Hembree, G. G., and Scheinfein, M. R., Self-organized Fe nanowire arrays prepared by shadow deposition on NaCl(110) template. *Appl. Phys. Lett.* **70,** 1043 (1997).

Suh, J. S., and Lee, J. S., Highly ordered two-dimensional carbon nanotube arrays. *Appl. Phys. Lett.* **75,** 2047 (1999).

Sun, L., Searson, P. C., and Chien, C. L., Electrochemical deposition of nickel nanowire arrays in single-crystal mica films. *Appl. Phys. Lett.* **74,** 2803 (1999a).

Sun, X., Zhang, Z., and Dresselhaus, M. S., Theoretical modeling of thermoelectricity in bismuth nanowires. *Appl. Phys. Lett.* **74,** 4005 (1999b).

Sun, L., Searson, P. C., and Chien, C. L., Finite-size effects in nickel nanowire arrays. *Phys. Rev. B* **61,** R6463 (2000).

Tang, Y. H., Wang, N., Zhang, Y. F., Lee, C. S., Bello, I., and Lee, S. T., Synthesis and characterization of amorphous carbon nanowires. *Appl. Phys. Lett.* **75,** 2921 (1999).

Tonucci, R. J., Justus, B. J., Campillo, A. J., and Ford, C. E., Nanochannel array glass. *Science* **258,** 783 (1992).

van Wees, B. J., van Houten, H., Beenakker, C. W. J., Williamson, J. G., Kouwenhoven, L. P., van der Marel, D., and Foxon, C. T., Quantized conductance of point contacts in a two-dimensional electron gas. *Phys. Rev. Lett.* **60,** 848 (1988).

Wagner, R. S., and Ellis, W. C., Nanowire formation in a polymeric film. *Appl. Phys. Lett.* **4,** 89 (1964).

Wang, N., Tang, Y. H., Zhang, Y. F., Lee, C. S., and Lee, S. T., Nucleation and growth of Si nanowires from silicon oxide. *Phys. Rev. B* **58,** R16024 (1998a).

Wang, N., Zhang, Y. F., Tang, Y. H., Lee, C. S., and Lee, S. T., SiO_2-enhanced synthesis of Si nanowires by laser ablation. *Appl. Phys. Lett.* **73,** 3902 (1998b).

Wharam, D. A., Thornton, T. J., Newbury, R., Pepper, M., Ahmed, H., Frost, J. E. F., Hasko, D. G., Peacock, D. C., Ritchie, D. A., and Jones, G. A. C., One-dimensional transport and the quantization of the ballistic resistance. *J. Phys. C* **21,** L209 (1988).

Whitney, T. M., Jiang, J. S., Searson, P. C., and Chien, C. L., Fabrication and magnetic-properties of arrays of metallic nanowires. *Science* **261,** 1316 (1993).

Wu, C.-G., and Bein, T., Conducting polyaniline filaments in a mesoporous channel host. *Science* **264,** 1757 (1994).

Xu, D., Chen, D., Xu, Y., Shi, X., Guo, G., Gui, L., and Tang, Y., Preparation of II–VI group semiconductor nanowire arrays by dc electrochemical deposition in porous aluminum oxide templates. *Pure Appl. Chem.* **72,** 127 (2000a).

Xu, D., Xu, Y., Chen, D., Guo, G., Gui, L., and Tang, Y., Preparation of CdS single-crystal nanowires by electrochemically induced deposition. *Adv. Mater.* **12,** 520 (2000b).

Yi, G., and Schwarzacher, W., Single crystal superconductor nanowires by electrodeposition. *Appl. Phys. Lett.* **74,** 1746 (1999).

Ying, J. Y., Nanoporous systems and templates: The unique self-assembly and synthesis of nanostructures. *Science Spectra* **18,** 56 (1999).

Ying, J. Y., Mehnert, C. P., and Wong, M. S., Synthesis and applications of supramolecular-templated mesoporous materials. *Angew. Chem. Int. Ed.* **38,** 56 (1999).

Yu, D. P., Lee, C. S., Bello, I., Zhou, G. W., and Bai, Z. G., Synthesis of nano-scale silicon wires by excimer laser ablation at high temperature. *Solid State Commun.* **105,** 403 (1998).

Zeng, H., Zheng, M., Skomski, R., Sellmyer, D. J., Liu, Y., Menon, L., and Bandyopadhyay, S., Magnetic properties of self-assembled Co nanowires of varying length and diameter. *J. Appl. Phys.* **87,** 4718 (2000).

Zhang, Z., Ying, J. Y., and Dresselhaus, M. S., Bismuth quantum-wire arrays fabricated by a vacuum melting and pressure injection process. *J. Mater. Res.* **13,** 1745 (1998a).

Zhang, Y. F., Tang, Y. H., Wang, N., Yu, D. P., Lee, C. S., Bello, I., and Lee, S. T., Silicon nanowires prepared by laser ablation at high temperature. *Appl. Phys. Lett.* **72,** 1835 (1998b).

Zhang, Z., Sun, X., Dresselhaus, M. S., Ying, J. Y., and Heremans, J., Magnetotransport investigation of ultrafine single-crystalline bismuth nanowire arrays. *Appl. Phys. Lett.* **73,** 1589 (1998c).

Zhang, Z., Gekhtman, D., Dresselhaus, M. S., and Ying, J. Y., Processing and characterization of single-crystalline ultrafine bismuth nanowires. *Chem. Mater.* **11,** 1659 (1999).

Zhang, Y. F., Tang, Y. H., Wang, N., Lee, C. S., Bello, I., and Lee, S. T., Synthesis and characterization of amorphous carbon nanowires. *Phys. Rev. B* **61,** 4518 (2000a).

Zhang, Z., Sun, X., Dresselhaus, M. S., Ying, J. Y., and Heremans, J., Electronic transport properties of single crystal bismuth nanowire arrays. *Phys. Rev. B* **61,** 4850 (2000b).

INDEX

A

Acetylene adsorption, selective, 117
Acoustic cavitation, nanostructured catalysts, 19
Activated alumina
- commercial, 93
- commercial use, 80
- pore size distribution, 89

Activated carbon
- adsorption of water vapor, 91
- commercial use, 80
- effects of oxidation on water adsorption, 91
- equilibrium sorption of water vapor, 90–91
- fibers, 91–92
- interests in hydrogen storage by adsorption, 92
- manufacture and use, 88
- methane, 92
- pore size distribution, 89
- raw materials, 88–89
- recent developments, 91–92
- tailoring pore structure for applications, 89–90
- unique surface property, 90

Adsorbents, *see* Nanostructured adsorbents
Adsorption
- effect of adsorbate properties, 84–85
- heat of, 83–84
- potential energies for, 81–83
- properties of pillared clays, 120
- properties of zeolites, 99–100
- threshold pressure in pore sizes and shapes, 87
- *see also* Nanostructured adsorbents

Advanced catalysts
- hydrodynamic cavitation, 3
- nanostructured materials, 2–3

Aerosol catalyst studies, 19
Ag, *see* Silver on alumina
Alumina
- implant material, 145
- *see also* Silver on alumina

Alumina templates
- anodic, 169–170
- processing anodic films, 170
- scanning electron microscopy (SEM) of porous, 171
- schematic, 170
- self-organized pore structure, 171

Antimony nanowires, temperature-dependent resistivity, 197–198

B

Ballistic transport, 191
Band structures, nanowire
- calculated subband energies as function of in-plane mass anisotropy, 188
- carrier densities, 190–191
- dispersion relation of electrons, 185
- envelope wavefunction of electrons, 186
- grid points transforming differential equation to difference equation, 187
- infinitely long wire with circular cross section, 186
- one-dimensional systems, 185–188
- parameters determining electronic properties, 186–187
- semimetal–semiconductor transition in semimetallic nanowires, 188–191
- subband energy, 187
- theoretical modeling, 185–191

Barium hexaaluminate, reverse microemulsion, 13–14
Bending properties, nanophase ceramics, 158
Bernoulli effect, cavitational bubble formation, 22
Bioceramics
- effect of surface properties, 146
- inconsistency of efficacy, 146
- orthopedic and dental implants, 145–146
- *see also* Nanophase ceramics

Bioglasses
implant material, 145
surface modification, 147–148
Biomaterials, *see* Orthopedic and dental implant materials
β-Bismuth molybdate, hydrodynamic cavitation, 33–34
Bismuth nanowires
calculated effective density of states, 189
carrier density, 190–191
density of states of electrons, 188
parameters determining electronic properties, 186–187
pressure injection method, 173–177
scanning electron microscopy (SEM) image of, on anodic alumina template, 175
smallest diameter, 177
temperature-dependent resistivity, 194–196
temperature-dependent resistivity of Bi–Te alloys, 196–197
transmission electron microscopy (TEM) image of cross section, 175, 176
X-ray diffraction (XRD) patterns, 176, 177
see also Nanowires
Bonding
π-complexation sorbents, 114–117
supported metal complexes, 62
supported metal nanoclusters, 73
Bone
bone-modeling unit (BMU), 139
cells of bone tissue, 136–139
chemical composition of matrix, 131–136
collagen, 132–133
cutter-cone, 139
fibroblasts, 138–139
fibronectin, 134–135
inorganic phase, 132
laminin, 133–134
mechanical properties, 128, 129
microarchitecture, 128, 130–131
microarchitecture and structural classification, 130
noncollagenous proteins, 133–136
organic phase, 132–136
osteoblasts, 136–138
osteoclasts, 138
physiology, 128–140
remodeling, 139–140
structural organization of microarchitecture, 131
time course of osteoblast function, 137
vitronectin, 135–136
wound-healing response, 141
see also Orthopedic and dental implant materials

C

Cabot flame process, high-surface-area nanostructured materials, 11
Calcium fluoride
evidence for shock wave heating, 31–32
Tamman temperature, 32
Carbon, *see* Activated carbon
Carbon nanotubes
derivatives of C_{60} buckyballs, 118–119
vapor deposition, 180–181
Carrier density, bismuth nanowires, 190–191
Catalysis, supported metal complexes, 62–63
Catalysts
connections of supported nanoclusters to industrial, 72
single-site, 51
supported metal nanoclusters, 73
supported nanostructures, 50–51
see also Nanostructured catalysts
Catalytic properties
nanostructured materials, 6–8
supported metal nanoclusters, 70–72
supported nanoparticles, 73
Cations
effects of, charge and ionic radius, 105–108
effects of, sites on zeolites, 102–105
interactions of zeolites with, 100–108
CaviMax processor
description, 26
experimental, 23–25
experiments, 26
see also Hydrodynamic cavitation
CaviPro processor
characterizing fluid flow, 26
description, 25
experimental, 23–25
flow configuration, 24
see also Hydrodynamic cavitation
Cavitation, *see* Hydrodynamic cavitation

Cell adhesion, protein-mediated, on biomaterials, 143–145
Ceramics
  hydrodynamic cavitation, 3
  orthopedic and dental implants, 145
  surface modification, 147–148
  see also Nanophase ceramics
Chemical composition of bone
  bone matrix, 131–136
  collagen in human body, 132–133
  fibronectin in bone matrix, 134–135
  inorganic phase, 132
  laminin in basement membrane of tissues, 133–134
  noncollagenous proteins in bone matrix, 133–136
  organic phase, 132–136
  vitronectin, 135–136
Chemical etching, porous templates, 171–172
Chemical modifications, implant materials, 147–148
Chemical vapor deposition, nanowires, 179
Cluster structures
  change in grain size, 5
  ion-bombardment technique, 7–8
Cobalt molybdates, hydrodesulfurization catalysts, 2–3
Collagen, human body, 132–133
Combustion flame–chemical vapor condensation (CF–CVC)
  nanostructured materials, 10–11
  schematic, 10
π-Complexation sorbents
  description, 108–109
  effects of cation, anion, and substrate, 112–113
  electron population changes on *d* orbitals, 116
  electron redistribution, 116–117
  energy of adsorption for MCl–C_2H_4 and MCl–CO systems, 112
  equilibrium isotherms of C_2H_4 over C_2H_6 on $AgNO_3/SiO_2$, 111
  equilibrium isotherms of C_3H_6 over C_3H_8 on $AgNO_3/SiO_2$, 111
  nature of π-complexation bond, 114–117
  normalized C_3H_6 adsorption isotherm on AgX salts, 113
  olefin–diene separation and purification, 117
  olefin–paraffin separations, 109–111
  schematic of metal–olefin complex, 115
  selective acetylene adsorption, 117
  separating aromatics from aliphatics, 117
  spontaneous monolayer dispersion, 110
  summary of natural bond orbital (NBO) analysis between MX and C_2H_4, 114
  summary of NBO analysis between MX and C_3H_6, 115
Composition, supported metal complexes, 53
Coordination chemistry, supported metal complexes, 52–53
Copper-modified zinc oxide
  grain sizes and crystallographic strain, 39
  hydrodynamic cavitation, 38–39
Crystallographic strain
  copper-modified zinc oxide, 38–39
  hydrodynamic cavitation, 34–39
  piezoelectrics, 37–38
  titania, 35–37
Cutter-cone, bone remodeling, 139

D

Decarbonylation, supported metal nanoclusters, 67
Dendrimers, polymer-mediated synthesis, 14–15
Density of states, bismuth nanowires, 188, 189
Dental implants, *see* Orthopedic and dental implant materials
Deprotonation of hydrido metal, supported metal nanoclusters, 66
Diffusive transport, 191
Dipole moment, effect on adsorption, 84–85
Dispersion
  contribution to potential energy for adsorption, 81–82
  Lennard–Jones potential, 83

E

Electrochemical deposition
  advantage, 179
  choosing template, 178
  contrast to pressure injection, 178–179
  nanowire fabrication, 177–179

Electronic charge, sorbent design, 85–86
Electronic materials, hydrodynamic cavitation, 3
Electronic properties, nanostructured materials, 4–5
Electron redistribution
phenomenon of *d* orbitals, 116–117
schematic, 117
Endothelial cells, protein-mediated cell adhesion on biomaterials, 143

F

Fibroblasts
formation of fibrous, connective tissue, 138–139
protein-mediated cell adhesion on biomaterials, 143
Fibronectin
glycoprotein, 134–135
schematic of structure, 135
Fluid-flow conditions
Ag on alumina, 41–42
high-temperature stable oxidation catalysts, 40–41
Pd on zirconia–alumina, 40–41
synthesis under variable, 39–42
variable phases under different bubble dynamics, 41–42

G

Gas phase condensation
dc- and rf-magnetron sputtering, 11
deposition of palladium on silicon carbide support, 12
nanostructured materials, 11–12
Geometry, sorbent design, 87–88
Glycoproteins
fibronectin, 134–135
integrins, 143–144
laminin, 133–134
vitronectin, 135–136
Gold particles
preparation on titania and iron oxide supports, 6–7
protein microtube-mediated synthesis, 16
Grain size
change resulting in cluster formation, 5
nanostructured metals and alloys in varying, 12
varying size of crystallites, 4

H

Heteropoly compounds, nanostructured adsorbents, 119
High-powered ultrasound
metal oxides and supported metal catalysts, 3
metal oxides and supported metal oxides, 17–18
High temperature aerosol decomposition, catalyst synthesis, 3
Hydrodynamic cavitation
advanced catalysts, ceramics, and electronic materials, 3
β-bismuth molybdate, 33–34
Bernoulli effect, 22
bubble formation-collapse-reformation dynamics, 22
calcium fluoride, 31–32
CaviMax and CaviPro processors, 23–24
copper-modified zinc oxide with crystallographic strain, 38–39
equipment, 21–22
experimental, 23–25
flow configuration in CaviPro, 24
high-temperature stable oxidation catalysts, 40–41
introduction of crystallographic strain, 34–39
$La_{0.6}Sr_{0.4}FeO_3$ perovskites, 25
MoO_3 synthesis, 28–32
nanostructured catalyst in high-phase purity and varying grain sizes, 32–34
piezoelectrics with crystallographic strain, 37–38
relationship of double-orifice cavitation generator to model, 23
Reynolds and throat cavitation numbers, 25–27
schematic, 21
schematic of cavitation bubbles interacting with slurry, 22

synthesis of metal oxide catalysts and supported metals, 27
technique, 20
titania synthesis, 25
titania with crystallographic strain, 35–37
typical experiment, 24–25
variable fluid-flow conditions, 39–42
variable phases under different bubble dynamics, 41–42
X-ray diffraction (XRD) experiments, 25
XRD of $La_{0.6}Sr_{0.4}FeO_3$, 33
Hydroxyapatite, implant material, 145

I

Industrial catalysts, connections of supported nanoclusters to, 72
Inorganic phase, bone, 132
Integrins
expression by osteoblasts, 144
function, 144
glycoproteins, 143–144
Ion-bombardment technique, metal cluster formation, 7–8
Ionic radii, alkali and alkaline earth metal ions, 87
$Ir_4(CO)_{12}$ and $Ir_6(CO)_{16}$, supported metal nanoclusters, 68–69
Ir_4 in zeolite NaX
supported metal nanoclusters, 69
theoretical investigation, 70
Iron oxide support, preparation of gold particles on, 6–7

K

Kirkwood–Muller formula, potential energies for adsorption, 83

L

Laminin
glycoproteins, 133–134
schematic of structure, 134
Laser-assisted synthesis
future research in laser ablation, 183–184
laser ablation-condensation, 181–182
nanowires, 181–184
schematic of silicon nanowires, 181
Lyotropic liquid-crystal template, polymer-mediated synthesis, 15

M

MCM-41
development by sol-gel route, 94
host matrices for nanowire fabrication, 172–173
mesoporous silicate/aluminosilicate, 94
modifications to be useful, 96
pore structure and volume, 95
promising applications, 96
schematic of formation, 95
synthesis and tailoring structure and properties, 95
Mechanical properties
bone, 128, 129
nanophase ceramics, 156–159
Metal alloys
implant material, 145
surface modification, 147–148
Metal clusters, ion-bombardment technique, 7–8
Metal complexes, *see* Supported metal complexes
Metal nanoclusters, *see* Supported metal nanoclusters
Metal–olefin complex, schematic, 115
Metal oxidation state, supported metal complexes, 53–54
Metal oxides
catalysts by hydrodynamic cavitation, 27
high-powered ultrasound, 3, 17–18
schematic of hydrodynamic cavitation synthesis, 21
Metal pair sites on supports, 63–64
Metalloorganic chemical vapor deposition, nanowires, 179
Metals
orthopedic and dental implants, 145
surface modification, 147–148
MgO, Os_5C on, 69
MgO supports, rhenium carbonyl precursors, 64

Mica film
fabrication, 172
scanning electron microscopy (SEM) image, 172
Microarchitecture
bone, 128, 130–131
structural organization of bone, 131
Microwave reactor, examining nanostructured catalyst synthesis, 19–20
Molecular-sieve carbon, pore size distribution, 89
Molecular-sieve zeolites, commercial use, 80
Molybdenum carbide, Mo_2C, sonochemical synthesis, 16–17
Molybdenum species, SiO_2-supported, 63
MoO_3 synthesis
calcium fluoride for evidence of shock wave heating, 31–32
CaviPro and CaviMax processors, 28–29
cavitational heating, 31
computed activation energy, 30
degree of in situ calcination, 28
reaction rate calculation, 30
sample from lowest temperature oven calcination, 29–30
shock wave, 30
XRD analysis of oven-calcined ammonium molybdate, 29
MoS_2
high edge-to-basal plane ratio, 6
sonochemical synthesis, 16

N

Nanochannel glass (NCG)
scanning electron microscopy (SEM), 173
schematic of fabrication, 173
template-assisted synthesis of nanowires, 172
Nanoparticles, *see* Supported nanoparticles
Nanophase ceramics
adhesion of osteoblasts to ceramic surfaces, 152–153
bending properties, 158
enhancing osteoblast and osteoclast functions, 153–155
experimental evidence on mechanical properties, 157–159
experimental evidence on surface properties, 150–156
extent of osteointegration of calcium phosphate with juxtaposed bone, 151
mechanical properties enhancing implant efficacy, 156–159
mechanism of osteoblast adhesion, 151–152
porosity changes, 158–159
rationale for mechanical properties, 156–157
rationale for surface properties, 149–150
representative topography of nanophase and conventional titania, 154
schematic deformation properties of, 157
studies on ectopic bone formation, 150
surface properties for enhanced implant efficacy, 149–156
theoretical predictions of changes in surface properties, 149
topography, 152
unfolding of vitronectin for osteoblast adhesion on, 155
see also Orthopedic and dental implant materials
Nanostructured adsorbents
activated alumina and silica gel, 93–94
activated carbon, 88–91
basic considerations for sorbent design, 85–88
carbon nanotubes, 118–119
commercial use of adsorption, 80
π-complexation sorbents, 108–117
contributions to initial heat of adsorption, 85
dipole moment, 84–85
dispersion, 82
dispersion, electrostatic, and chemical bond contributions, 81–82
effects of adsorbate properties on adsorption, 84–85
factors for designing adsorbents, 81–88
field and induced-point dipole, 82
field and point dipole, 83
field gradient and linear point quadrupole, 83
heat of adsorption, 83–84
heteropoly compounds, 119
individual contributions, 82
ionic radii, 87

Kirkwood–Muller formula, 83
Lennard–Jones potential, 83
nonspecific contributions, 82
pillared clays, 120
polarizabilities of ground-state atoms and ions, 86
polarizability, 84–85
polarizability, electronic charge, and van der Waals radius, 85–86
pore size and geometry, 87–88
potential energies for adsorption, 81–83
quadrupole moment, 84–85
recent developments on activated carbon, 91–92
repulsion, 82
threshold pressure for adsorption in different pore sizes and shapes, 87
zeolites, 96–108
see also Activated carbon; π-Complexation sorbents; Zeolites

Nanostructured catalysts
acoustic cavitation, 19
aerosol catalyst studies, 19
β-bismuth molybdate and perovskite $La_{0.6}Sr_{0.4}FeO_3$, 33–34
β-bismuth molybdate using both CaviMax processors, 34
CaviMax and CaviPro processors, 23–26
cavitational devices, 23–24
characterization of Reynolds and throat cavitation numbers, 25–27
classical approach, 18
copper-modified zinc oxide with crystallographic strain, 38–39
crystallographic strain by hydrodynamic cavitation, 34–39
engineered synthesis, 18–19
equipment for hydrodynamic cavitation, 21–22
estimating in situ calcination temperature in MoO_3 synthesis, 28–32
experimental procedures, 23–25
high-temperature stable oxidation catalysts, 40–41
hydrodynamic cavitation, 20–23
hydrodynamic cavitation with high-phase purities and varying grain sizes, 32–34
mechanical techniques affording changes, 18–19
metal oxide catalysts and supported metals by hydrodynamic cavitation, 27
microwave reactor examining synthesis, 19–20
piezoelectrics with crystallographic strain, 37–38
schematic of hydrodynamic cavitation, 21
throat cavitation number, 26
titania with crystallographic strain, 35–37
typical experiment, 24–25
variable fluid-flow conditions, 39–42
variable phases under different bubble dynamics, 41–42
X-ray diffraction (XRD) experiments, 25
see also MoO_3 synthesis

Nanostructured materials
barium hexaaluminate by reverse microemulsion, 13–14
Cabot flame process, 11
catalytic properties, 6–8
combustion flame–chemical vapor condensation process (CF–CVC), 10–11
factors responsible for rate acceleration, 6
formation of cluster structures, 5
gas phase condensation synthesis, 11–12
high metal atom surface-to-volume ratio, 4–5
high-powered ultrasound, 3, 17–18
high-temperature aerosol decomposition (HTAD) process, 3
hydrodesulfurization catalysts, 2–3
hydrodynamic cavitation, 3
ion-bombardment technique, 7
ionization potential, 4
MoS_2 with high edge-to-basal plane ratio, 6
new synthesis processes, 2
palladium on silicon carbide support, 12
polymer-mediated synthesis, 14–15
preparation of gold particles on titania and iron oxide, 6–7
progress in synthesis processes, 8–9
properties and reactivities, 3–8
protein microtubule-mediated synthesis, 15–16
reverse micelle synthesis, 12–14
sol-gel and precipitation technologies, 9–10
sonochemical synthesis, 16–18
structure and electronic properties, 4–5

Nanostructured materials (*continued*)
synthesis for advanced catalysts, 2–3
synthesis of TiO_2 by aerosol process, 6
varying grain size of crystallites, 4
Nanowires
advances in low-dimensional systems, 198
anodic alumina templates, 169–170
antimony nanowires, 197–198
ballistic transport, 191
band structures of one-dimensional systems, 185–188
bismuth, 175–177
calculated effective densities of states for bismuth nanowires, 189
calculated subband energies as function of in-plane mass anisotropy, 188
carbon nanotubes, 180–181
carrier densities, 190–191
diffusive transport, 191
dispersion relation of electrons, 185
electrochemical deposition, 177–179
extension beyond gas-phase reaction for Si nanowires, 184
further research in laser ablation, 183–184
grid points transforming differential equation into difference equation, 187
laser ablation-condensation, 181–182
laser-assisted synthesis, 181–184
mica films, 172
molecular sieve MCM-41, 172–173
nanochannel glass (NCG), 172
parameters determining electronic properties, 186–187
physical vapor deposition, 179–180
pressure injection method, 173–177
processing of anodic alumina films, 170
progress in deriving, 168–169
resistance R(T)/R(270 K) for Bi–Te nanowire alloys, 196–197
schematic of pressure injection setup, 174
schematic of semimetal–semiconductor transition, 189
self-organized pore structure of anodic alumina, 171
semiclassical model of transport, 192–193
semimetal–semiconductor transition in semimetallic, 188–191
silicon nanowires, 181–183
techniques to prepare one-dimensional systems, 184–185
temperature dependence of resistance for bismuth nanowire arrays, 194–196
temperature-dependent resistivity, 193–198
template-assisted synthesis, 169–181
template fabrication by chemical etching, 171–172
theoretical modeling of band structures, 185–191
transport properties, 191–198
vapor deposition, 179–181
vapor–liquid–solid (VLS) growth, 181
Natural bond orbital (NBO), π-complexation sorbents, 114–115
Noncollagenous proteins
bone matrix, 133–136
fibronectin, 134–135
laminin, 133–134
vitronectin, 135–136

O

Olefin–paraffin separation, π-complexation sorbents, 109–111
Organic phase
bone, 132–136
collagen, 132–133
fibronectin, 134–135
laminin, 133–134
noncollagenous materials, 133–136
vitronectin, 135–136
Organometallic chemistry, supported metal complexes, 52
Orthopedic and dental implant materials
bioceramics, 145–146
chemical modifications, 147–148
comparing mechanical properties of, and bone, 146
conventional, 127
costs, 126–127
current materials, 145–148
fate of implanted device, 140–141
integration into surrounding tissue, 127
integrin expression on osteoblasts, 144
integrins, 143–144
metals, ceramics, and polymers, 145
next generation, 127, 148–159

novel surface modifications of conventional, 147–148
peptide sequence tyrosine-arginine-serine-arginine (KRSR), 144–145
protein interactions with biomaterial surfaces, 141–143
protein-mediated cell adhesion on biomaterials, 143–145
schematic of protein-mediated cell adhesion on biomaterials, 142
surface roughness, 147
tissue–implant interface, 140–145
use, 126–127
wound-healing response of bone, 141
see also Nanophase ceramics

Os_5C on MgO, supported metal nanoclusters, 69
Osmium di- and tricarbonyls on γ-Al_2O_3, supported metal complexes, 61
Osteoblasts
adhesion to ceramic surfaces, 152–153
bone-forming cells, 136–138
bone remodeling, 139–140
integrin expression by, 144
mechanism of vitronectin mediating, adhesion, 151–152
nanoceramics enhancing, 153–155
periods of differentiation, 136, 137
protein-mediated cell adhesion on biomaterials, 143
surface modifications of implant materials, 147–148
unfolding of vitronectin for, adhesion, 155
Osteoclasts
bone remodeling, 139–140
bone-resorbing cells, 138
nanoceramics enhancing, 153–155

P

Palladium
dendrimers, 14–15
protein microtube-mediated synthesis, 15–16
silicon carbide support by plasma-sputtering, 12
synthesis on zirconia–alumina, 40–41
X-ray diffraction for 2% Pd on zirconia–alumina, 40
Perovskite $La_{0.6}Sr_{0.4}FeO_3$
hydrodynamic cavitation, 33–34
synthesis of family, 25
Physical vapor deposition, nanowires, 179–180
Physiology, *see* Bone
Piezoelectrics, hydrodynamic cavitation, 37–38
Pillared interlayered clays (PILCs), nanostructured adsorbents, 120
Platinum, dendrimers, 14–15
Polarizability
effect on adsorption, 84–85
ground-state atoms and ions, 86
sorbent design, 85–86
Polycarbonate membrane
chemical etching, 171–172
scanning electron microscopy of particle track-etched, 172
Polyethylene, implant material, 145
Polymer-mediated synthesis, dendrimers, 14–15
Polymers, orthopedic and dental implants, 145
Poly(methyl methacrylate), implant material, 145
Pore size, sorbent design, 87–88
Potential energy, adsorption, 81–83
Precipitation, nanostructured materials, 9–10
Pressure injection
bismuth nanowires, 175–177
experimental setup, 174
nanowire fabrication, 173–177
template requirements, 175
Washburn equation, 174–175
Pressure swing adsorption, adsorption, 80
Protein microtube-mediated synthesis, nanostructured materials, 15–16
Purification, olefin–diene, 117

Q

Quadrupole moment, effect on adsorption, 84–85

R

Reactivity
  supported metal complexes, 62–63
  supported metal nanoclusters, 73
Repulsion
  contribution to potential energies for adsorption, 82
  Lennard–Jones potential, 83
Reverse micelle process
  combining with standard sol-gel, 13
  inverse micelle synthesis of nanostructured reduced metals, 13
  nanostructured materials, 12–14
  silver on fine-grain silica, 13
Reynolds number, characterization, 25–27
$Rh_6(CO)_{16}$, supported metal nanoclusters, 68
Rhenium carbonyl precursors, MgO supports, 64
Rhenium tricarbonyl on MgO
  preparation, 58
  structural assignments and infrared spectra, 59
  supported metal complexes, 58–61
  theoretical and experimental agreement, 60–61
Rhodium dicarbonyls on dealuminated Y zeolite
  EXAFS spectra and predictions of density functional theory, 57
  infrared spectra in carbonyl stretching region, 56
  model, 55
  supported metal complexes, 54–56
Rhodium dicarbonyls on γ-Al_2O_3, supported metal complexes, 58

S

Semimetallic nanowires
  carrier density, 190–191
  schematic of transition, 189
  semimetal–semiconductor transition, 188–191
Separations
  aromatic from aliphatic, 117
  olefin–diene, 117
Silica gel
  amorphous, 93
  commercial, 93
  commercial use, 80
  pore size distribution, 89
  sol-gel processing, 93–94
Silicon carbide support, palladium deposition by plasma-sputtering, 12
Silicon nanowires
  schematic, 181
  transmission electron microscopy (TEM) images, 182–183
  vapor–liquid–solid (VLS) growth, 181, 184
Silver on alumina
  variable phases under different bubble dynamics, 41–42
  X-ray diffraction patterns, 41
SiO_2 supports
  molybdenum species, 63
  tungsten species, 63
Sol-gel
  advantages, 9
  combining with inverse micelle technique, 13
  development of MCM-41, 94
  nanostructured materials, 9–10
  silica gel, 93–94
Sonochemical synthesis
  metallic and bimetallic catalysts, 16
  molybdenum carbide (Mo_2C), 16–17
  MoS_2, 16
  nanostructured materials, 16–18
Sorbent design
  polarizability, electronic charge, and van der Waals radius, 85–86
  pore size and geometry, 87–88
Spontaneous monolayer dispersion, π-complexation sorbents, 110
Structure
  characterization for supported metal complexes, 54
  characterization for supported metal nanoclusters, 67
  nanostructured materials, 4–5
  supported metal complexes, 62
  supported metal nanoclusters, 73
  zeolites, 97, 98–99
Subband energies
  bismuth nanowires, 186–187
  calculated, as function of in-plane mass anisotropy, 188
Supported metal complexes

composition determination, 53
coordination chemistry, 52–53
examples, 54–62
metal oxidation state determination, 53–54
organometallic chemistry, 52
osmium di- and tricarbonyls on γ-Al_2O_3, 61
preparation, 52–53
reactivity and catalysis, 62–63
rhenium tricarbonyls on MgO, 58–61
rhodium dicarbonyls on dealuminated Y zeolite, 54–56
rhodium dicarbonyls on γ-Al_2O_3, 58
single-site catalysts, 51
spectroscopic and theoretical characterization of structure, 54
structure and bonding, 62
tantalum hydride complexes on SiO_2, 62
Supported metal nanoclusters
catalytic activities of extremely small clusters, 72
catalytic properties, 70–72
cluster-size dependence, 72
connections to industrial catalysts, 72
decarbonylation of neutral or anionic metal carbonyl clusters, 67
deprotonation of hydrido metal carbonyl cluster, 66
EXAFS results characterizing, 71
examples, 68–70
Ir_4 in zeolite NaX, 69, 70
$[Ir_4(CO)_{12}]$ and $[Ir_6(CO)_{16}]$, 68–69
metal carbonyl clusters, 66
Os_5C on MgO, 69
preparation, 65–67
$[Rh_6(CO)_{16}]$, 68
solution reactions in presence of reducing agents, 66–67
structural characterization, 67
structure, bonding, reactivity, and catalysis, 73
surface chemistry, 66
syntheses in zeolite cages, 67
theoretical metal–metal distances, 70
Supported metals, hydrodynamic cavitation, 27
Supported nanoparticles, properties, 73
Surface chemistry, supported metal nanoclusters, 66
Surface properties, nanophase ceramics, 149–156
Surface roughness, modifications of implant materials, 147
Surface-to-volume ratio, metal atom, 4–5

T

Tantalum hydride on SiO_2, supported metal complexes, 62
Template-assisted synthesis
anodic alumina templates, 169–170
description, 169
electrochemical deposition, 177–179
nanowires, 169–181
pressure injection, 173–177
vapor deposition, 179–181
Throat cavitation number
characterization, 25–27
definition, 26
TiO_2, aerosol process, 6
Tissue–implant interface
fate of implanted device, 140–141
protein interactions with biomaterial surfaces, 141–143
protein-mediated cell adhesion on biomaterials, 143–145
wound-healing response of bone, 141
see also Orthopedic and dental implant materials
Titania
correlation of applied pressure, crystallographic strain, and Reynolds number–throat cavitation number, 36
grain size and crystallographic strain data, 35
hydrodynamic cavitation, 35–37
preparation of gold particles on, 6–7
synthesis of nanostructured, 9–10
topography of nanophase and conventional, 154
Topography
nanophase and conventional titania, 154
nanophase ceramics, 152
Transport properties
antimony nanowires, 197–198
bismuth nanowires, 194–196
Bi–Te alloys, 196–197
reduced chemical potential, 193

Transport properties (*continued*)
semiclassical model, 192–193
temperature-dependent resistivity of nanowires, 193–198
Triplet sites on supports, 63–64
Tungsten species, SiO_2-supported, 63
Turnover numbers (TON), nanostructured materials, 6
Tyrosine-arginine-serine-arginine (KRSR), enhancing osteoblast adhesion, 144–145

V

van der Waals radius, sorbent design, 85–86
Vapor deposition
carbon nanotubes, 180–181
nanowire preparation, 179–181
Vapor–liquid–solid (VLS) growth, silicon nanowires, 181, 184
Vapor-phase synthesis, oxygen nonstoichiometry, 11
Vitronectin
bone matrix, 135–136
mechanism mediating osteoblast adhesion, 151–152
schematic of structure, 135
unfolding for osteoblast adhesion, 155

W

Washburn equation, pressure injection process, 174–175
Wound healing, response of bone, 141

Z

Zeolite cages, supported metal nanoclusters, 67
Zeolites
adsorption, 80
adsorption properties, 99–100
anionic oxygens and isolated cations, 99–100
components of interaction energies for CO_2 adsorbed on X zeolite, 107
description, 96–97
effects of cation charge and ionic radius, 105–108
effects of cation sites, 102–105
electrostatic interactions, 105–106
energy terms for NH_3 on X zeolite, 107
geometry-optimized cluster model, 101
heat of adsorption versus surface coverage, 105
interaction energies between molecules and isolated cations, 106
interaction potentials for CO_2 and NH_3, 107–108
interactions with cations, 100–108
Ir_4 in zeolite NaX, 69, 70
line representation of structure, 97
N_2 adsorption isotherms for Ag/LSX (low silica X), 103
N_2 and O_2 adsorption capacities for Li faujasite, 103
naturally occurring, 97–98
pore size distribution, 89
relative electronegativities of zeolite anion and halides, 101
site II cation on six-membered oxygen ring, 101
sodalite and hexagonal prism site I and II cation locations in Ag/faujasites, 104
structures and cation sites, 98–99
type A, 98–99
types X and Y, 99
Zinc oxide, copper modified, 38–39

CONTENTS OF VOLUMES IN THIS SERIAL

Volume 1

J. W. Westwater, *Boiling of Liquids*
A. B. Metzner, *Non-Newtonian Technology: Fluid Mechanics, Mixing, and Heat Transfer*
R. Byron Bird, *Theory of Diffusion*
J. B. Opfell and B. H. Sage, *Turbulence in Thermal and Material Transport*
Robert E. Treybal, *Mechanically Aided Liquid Extraction*
Robert W. Schrage, *The Automatic Computer in the Control and Planning of Manufacturing Operations*
Ernest J. Henley and Nathaniel F. Barr, *Ionizing Radiation Applied to Chemical Processes and to Food and Drug Processing*

Volume 2

J. W. Westwater, *Boiling of Liquids*
Ernest F. Johnson, *Automatic Process Control*
Bernard Manowitz, *Treatment and Disposal of Wastes in Nuclear Chemical Technology*
George A. Sofer and Harold C. Weingartner, *High Vacuum Technology*
Theodore Vermeulen, *Separation by Adsorption Methods*
Sherman S. Weidenbaum, *Mixing of Solids*

Volume 3

C. S. Grove, Jr., Robert V. Jelinek, and Herbert M. Schoen, *Crystallization from Solution*
F. Alan Ferguson and Russell C. Phillips, *High Temperature Technology*
Daniel Hyman, *Mixing and Agitation*
John Beck, *Design of Packed Catalytic Reactors*
Douglass J. Wilde, *Optimization Methods*

Volume 4

J. T. Davies, *Mass-Transfer and Interfacial Phenomena*
R. C. Kintner, *Drop Phenomena Affecting Liquid Extraction*
Octave Levenspiel and Kenneth B. Bischoff, *Patterns of Flow in Chemical Process Vessels*
Donald S. Scott, *Properties of Concurrent Gas–Liquid Flow*
D. N. Hanson and G. F. Somerville, *A General Program for Computing Multistage Vapor–Liquid Processes*

Volume 5

J. F. Wehner, *Flame Processes–Theoretical and Experimental*
J. H. Sinfelt, *Bifunctional Catalysts*
S. G. Bankoff, *Heat Conduction or Diffusion with Change of Phase*

George D. Fulford, *The Flow of Liquids in Thin Films*
K. Rietema, *Segregation in Liquid–Liquid Dispersions and Its Effect on Chemical Reactions*

Volume 6

S. G. Bankoff, *Diffusion-Controlled Bubble Growth*
John C. Berg, Andreas Acrivos, and Michel Boudart, *Evaporation Convection*
H. M. Tsuchiya, A. G. Fredrickson, and R. Aris, *Dynamics of Microbial Cell Populations*
Samuel Sideman, *Direct Contact Heat Transfer between Immiscible Liquids*
Howard Brenner, *Hydrodynamic Resistance of Particles at Small Reynolds Numbers*

Volume 7

Robert S. Brown, Ralph Anderson, and Larry J. Shannon, *Ignition and Combustion of Solid Rocket Propellants*
Knud Østergaard, *Gas–Liquid–Particle Operations in Chemical Reaction Engineering*
J. M. Prausnitz, *Thermodynamics of Fluid–Phase Equilibria at High Pressures*
Robert V. Macbeth, *The Burn–Out Phenomenon in Forced-Convection Boiling*
William Resnick and Benjamin Gal-Or, *Gas–Liquid Dispersions*

Volume 8

C. E. Lapple, *Electrostatic Phenomena with Particulates*
J. R. Kittrell, *Mathematical Modeling of Chemical Reactions*
W. P. Ledet and D. M. Himmelblau, *Decomposition Procedures for the Solving of Large Scale Systems*
R. Kumar and N. R. Kuloor, *The Formation of Bubbles and Drops*

Volume 9

Renato G. Bautista, *Hydrometallurgy*
Kishan B. Mathur and Norman Epstein, *Dynamics of Spouted Beds*
W. C. Reynolds, *Recent Advances in the Computation of Turbulent Flows*
R. E. Peck and D. T. Wasan, *Drying of Solid Particles and Sheets*

Volume 10

G. E. O'Connor and T. W. F. Russell, *Heat Transfer in Tubular Fluid–Fluid Systems*
P. C. Kapur, *Balling and Granulation*
Richard S. H. Mah and Mordechai Shacham, *Pipeline Network Design and Synthesis*
J. Robert Selman and Charles W. Tobias, *Mass-Transfer Measurements by the Limiting-Current Technique*

Volume 11

Jean-Claude Charpentier, *Mass-Transfer Rates in Gas–Liquid Absorbers and Reactors*
Dee H. Barker and C. R. Mitra, *The Indian Chemical Industry–Its Development and Needs*
Lawrence L. Tavlarides and Michael Stamatoudis, *The Analysis of Interphase Reactions and Mass Transfer in Liquid–Liquid Dispersions*
Terukatsu Miyauchi, Shintaro Furusaki, Shigeharu Morooka, and Yoneichi Ikeda, *Transport Phenomena and Reaction in Fluidized Catalyst Beds*

Volume 12

C. D. Prater, J. Wei, V. W. Weekman, Jr., and B. Gross, *A Reaction Engineering Case History: Coke Burning in Thermofor Catalytic Cracking Regenerators*
Costel D. Denson, *Stripping Operations in Polymer Processing*
Robert C. Reid, *Rapid Phase Transitions from Liquid to Vapor*
John H. Seinfeld, *Atmospheric Diffusion Theory*

Volume 13

Edward G. Jefferson, *Future Opportunities in Chemical Engineering*
Eli Ruckenstein, *Analysis of Transport Phenomena Using Scaling and Physical Models*
Rohit Khanna and John H. Seinfeld, *Mathematical Modeling of Packed Bed Reactors: Numerical Solutions and Control Model Development*
Michael P. Ramage, Kenneth R. Graziano, Paul H. Schipper, Frederick J. Krambeck, and Byung C. Choi, *KINPTR (Mobil's Kinetic Reforming Model): A Review of Mobil's Industrial Process Modeling Philosophy*

Volume 14

Richard D. Colberg and Manfred Morari, *Analysis and Synthesis of Resilient Heat Exchanger Networks*
Richard J. Quann, Robert A. Ware, Chi-Wen Hung, and James Wei, *Catalytic Hydrometallation of Petroleum*
Kent David, *The Safety Matrix: People Applying Technology to Yield Safe Chemical Plants and Products*

Volume 15

Pierre M. Adler, Ali Nadim, and Howard Brenner, *Rheological Models of Suspensions*
Stanley M. Englund, *Opportunities in the Design of Inherently Safer Chemical Plants*
H. J. Ploehn and W. B. Russel, *Interations between Colloidal Particles and Soluble Polymers*

Volume 16

Perspectives in Chemical Engineering: Research and Education

Clark K. Colton, *Editor*

Historical Perspective and Overview

L. E. Scriven, *On the Emergence and Evolution of Chemical Engineering*
Ralph Landau, *Academic–Industrial Interaction in the Early Development of Chemical Engineering*
James Wei, *Future Directions of Chemical Engineering*

Fluid Mechanics and Transport

L. G. Leal, *Challenges and Opportunities in Fluid Mechanics and Transport Phenomena*
William B. Russel, *Fluid Mechanics and Transport Research in Chemical Engineering*
J. R. A. Pearson, *Fluid Mechanics and Transport Phenomena*

Thermodynamics

Keith E. Gubbins, *Thermodynamics*
J. M. Prausnitz, *Chemical Engineering Thermodynamics: Continuity and Expanding Frontiers*
H. Ted Davis, *Future Opportunities in Thermodynamics*

Kinetics, Catalysis, and Reactor Engineering

Alexis T. Bell, *Reflections on the Current Status and Future Directions of Chemical Reaction Engineering*
James R. Katzer and S. S. Wong, *Frontiers in Chemical Reaction Engineering*
L. Louis Hegedus, *Catalyst Design*

Environmental Protection and Energy

John H. Seinfeld, *Environmental Chemical Engineering*
T. W. F. Russell, *Energy and Environmental Concerns*
Janos M. Beer, Jack B. Howard, John P. Longwell, and Adel F. Sarofim, *The Role of Chemical Engineering in Fuel Manufacture and Use of Fuels*

Polymers

Matthew Tirrell, *Polymer Science in Chemical Engineering*
Richard A. Register and Stuart L. Cooper, *Chemical Engineers in Polymer Science: The Need for an Interdisciplinary Approach*

Microelectronic and Optical Materials

Larry F. Thompson, *Chemical Engineering Research Opportunities in Electronic and Optical Materials Research*
Klavs F. Jensen, *Chemical Engineering in the Processing of Electronic and Optical Materials: A Discussion*

Bioengineering

James E. Bailey, *Bioprocess Engineering*
Arthur E. Humphrey, *Some Unsolved Problems of Biotechnology*
Channing Robertson, *Chemical Engineering: Its Role in the Medical and Health Sciences*

Process Engineering

Arthur W. Westerberg, *Process Engineering*
Manfred Morari, *Process Control Theory: Reflections on the Past Decade and Goals for the Next*
James M. Douglas, *The Paradigm After Next*
George Stephanopoulos, *Symbolic Computing and Artificial Intelligence in Chemical Engineering: A New Challenge*

The Identity of Our Profession

Morton M. Denn, *The Identity of Our Profession*

Volume 17

Y. T. Shah, *Design Parameters for Mechanically Agitated Reactors*
Mooson Kwauk, *Particulate Fluidization: An Overview*

Volume 18

E. James Davis, *Microchemical Engineering: The Physics and Chemistry of the Microparticle*
Selim M. Senkan, *Detailed Chemical Kinetic Modeling: Chemical Reaction Engineering of the Future*
Lorenz T. Biegler, *Optimization Strategies for Complex Process Models*

Volume 19

Robert Langer, *Polymer Systems for Controlled Release of Macromolecules, Immobilized Enzyme Medical Bioreactors, and Tissue Engineering*
J. J. Linderman, P. A. Mahama, K. E. Forsten, and D. A. Lauffenburger, *Diffusion and Probability in Receptor Binding and Signaling*
Rakesh K. Jain, *Transport Phenomena in Tumors*
R. Krishna, *A Systems Approach to Multiphase Reactor Selection*
David T. Allen, *Pollution Prevention: Engineering Design at Macro-, Meso-, and Microscales*
John H. Seinfeld, Jean. M. Andino, Frank M. Bowman, Hali J. L. Forstner, and Spyros Pandis, *Tropospheric Chemistry*

Volume 20

Arthur M. Squires, *Origins of the Fast Fluid Bed*
Yu Zhiqing, *Application Collocation*
Youchu Li, *Hydrodynamics*
Li Jinghai, *Modeling*
Yu Zhiqing and Jin Yong, *Heat and Mass Transfer*
Mooson Kwauk, *Powder Assessment*
Li Hongzhong, *Hardware Development*
Youchu Li and Xuyi Zhang, *Circulating Fluidized Bed Combustion*
Chen Junwu, Cao Hanchang, and Liu Taiji, *Catalyst Regeneration in Fluid Catalytic Cracking*

Volume 21

Christopher J. Nagel, Chonghun Han, and George Stephanopoulos, *Modeling Languages: Declarative and Imperative Descriptions of Chemical Reactions and Processing Systems*
Chonghun Han, George Stephanopoulos, and James M. Douglas, *Automation in Design: The Conceptual Synthesis of Chemical Processing Schemes*
Michael L. Mavrovouniotis, *Symbolic and Quantitative Reasoning: Design of Reaction Pathways through Recursive Satisfaction of Constraints*
Christopher Nagel and George Stephanopoulos, *Inductive and Deductive Reasoning: The Case of Identifying Potential Hazards in Chemical Processes*
Keven G. Joback and George Stephanopoulos, *Searching Spaces of Discrete Solutions: The Design of Molecules Processing Desired Physical Properties*

Volume 22

Chonghun Han, Ramachandran Lakshmanan, Bhavik Bakshi, and George Stephanopoulos, *Nonmonotonic Reasoning: The Synthesis of Operating Procedures in Chemical Plants*
Pedro M. Saraiva, *Inductive and Analogical Learning: Data-Driven Improvement of Process Operations*

Alexandros Koulouris, Bhavik R. Bakshi and George Stephanopoulos, *Empirical Learning through Neural Networks: The Wave-Net Solution*
Bhavik R. Bakshi and George Stephanopoulos, *Reasoning in Time: Modeling, Analysis, and Pattern Recognition of Temporal Process Trends*
Matthew J. Realff, *Intelligence in Numerical Computing: Improving Batch Scheduling Algorithms through Explanation-Based Learning*

Volume 23

Jeffrey J. Siirola, *Industrial Applications of Chemical Process Synthesis*
Arthur W. Westerberg and Oliver Wahnschafft, *The Synthesis of Distillation-Based Separation Systems*
Ignacio E. Grossmann, *Mixed-Integer Optimization Techniques for Algorithmic Process Synthesis*
Subash Balakrishna and Lorenz T. Biegler, *Chemical Reactor Network Targeting and Integration: An Optimization Approach*
Steve Walsh and John Perkins, *Operability and Control in Process Synthesis and Design*

Volume 24

Raffaella Ocone and Gianni Astarita, *Kinetics and Thermodynamics in Multicomponent Mixtures*
Arvind Varma, Alexander S. Rogachev, Alexandra S. Mukasyan, and Stephen Hwang, *Combustion Synthesis of Advanced Materials: Principles and Applications*
J. A. M. Kuipers and W. P. M. van Swaaij, *Computational Fluid Dynamics Applied to Chemical Reaction Engineering*
Ronald E. Schmitt, Howard Klee, Debora M. Sparks, and Mahesh K. Podar, *Using Relative Risk Analysis to Set Priorities for Pollution Prevention at a Petroleum Refinery*

Volume 25

J. F. Davis, M. J. Piovoso, K. A. Hoo, and B. R. Bakshi, *Process Data Analysis and Interpretation*
J. M. Ottino, P. DeRoussel, S. Hansen, and D. V. Khakhar, *Mixing and Dispersion of Viscous Liquids and Powdered Solids*
Peter L. Silveston, Li Chengyue, Yuan Wei-Kang, *Application of Periodic Operation to Sulfur Dioxide Oxidation*

Volume 26

J. B. Joshi, N. S. Deshpande, M. Dinkar, and D. V. Phanikumar, *Hydrodynamic Stability of Multiphase Reactors*
Michael Nikolaou, *Model Predictive Controllers: A Critical Synthesis of Theory and Industrial Needs*

Volume 27

William R. Moser, Josef Find, Sean C. Emerson, and Ivo M. Krausz, *Engineered Synthesis of Nanostructured Materials and Catalysts*
Bruce C. Gates, *Supported Nanostructured Catalysts: Metal Complexes and Metal Clusters*
Ralph T. Yang, *Nanostructured Adsorbents*
Thomas J. Webster, *Nanophase Ceramics: The Future Orthopaedic/Dental Implant Material*
Yu-Ming Lin, Mildred S. Dresselhaus, and Jackie Y. Ying, *Fabrication, Structure and Transport Properties of Nanowires*